ADVANCES IN CHEMISTRY RESEARCH

ADVANCES IN CHEMISTRY RESEARCH

VOLUME 35

ADVANCES IN CHEMISTRY RESEARCH

Additional books in this series can be found on Nova's website
under the Series tab.

Additional e-books in this series can be found on Nova's website
under the e-book tab.

ADVANCES IN CHEMISTRY RESEARCH

VOLUME 35

JAMES C. TAYLOR
EDITOR

Copyright © 2017 by Nova Science Publishers, Inc.

All rights reserved. No part of this book may be reproduced, stored in a retrieval system or transmitted in any form or by any means: electronic, electrostatic, magnetic, tape, mechanical photocopying, recording or otherwise without the written permission of the Publisher.

We have partnered with Copyright Clearance Center to make it easy for you to obtain permissions to reuse content from this publication. Simply navigate to this publication's page on Nova's website and locate the "Get Permission" button below the title description. This button is linked directly to the title's permission page on copyright.com. Alternatively, you can visit copyright.com and search by title, ISBN, or ISSN.

For further questions about using the service on copyright.com, please contact:
Copyright Clearance Center
Phone: +1-(978) 750-8400   Fax: +1-(978) 750-4470   E-mail: info@copyright.com.

NOTICE TO THE READER

The Publisher has taken reasonable care in the preparation of this book, but makes no expressed or implied warranty of any kind and assumes no responsibility for any errors or omissions. No liability is assumed for incidental or consequential damages in connection with or arising out of information contained in this book. The Publisher shall not be liable for any special, consequential, or exemplary damages resulting, in whole or in part, from the readers' use of, or reliance upon, this material. Any parts of this book based on government reports are so indicated and copyright is claimed for those parts to the extent applicable to compilations of such works.

Independent verification should be sought for any data, advice or recommendations contained in this book. In addition, no responsibility is assumed by the publisher for any injury and/or damage to persons or property arising from any methods, products, instructions, ideas or otherwise contained in this publication.

This publication is designed to provide accurate and authoritative information with regard to the subject matter covered herein. It is sold with the clear understanding that the Publisher is not engaged in rendering legal or any other professional services. If legal or any other expert assistance is required, the services of a competent person should be sought. FROM A DECLARATION OF PARTICIPANTS JOINTLY ADOPTED BY A COMMITTEE OF THE AMERICAN BAR ASSOCIATION AND A COMMITTEE OF PUBLISHERS.

Additional color graphics may be available in the e-book version of this book.

Library of Congress Cataloging-in-Publication Data

ISSN: 1940-0950
ISBN: 978-1-53610-734-0

Published by Nova Science Publishers, Inc. † New York

CONTENTS

Preface vii

Chapter 1 Recent Advances and Development of
Gliadin-Based Drug Delivery System 1
*Andresa da Costa Ribeiro, Daiani Canabarro Leite,
Rosane Michele Duarte Soares and
Nádya Pesce da Silveira*

Chapter 2 Lewis Acid Metal Salts-Catalyzed Synthesis of
β-Citronellol Derivatives 19
Márcio José da Silva

Chapter 3 Development and Applications of a New Type
of Polymer-Supported Organosilica
Layered-Hybrid Membranes 45
Genghao Gong and Toshinori Tsuru

Chapter 4 The Long-Range Order in Vegetable and Perfume
Oils, Butter and Animal Fat 61
*Kristina Zubow, Anatolij Zubow and
Viktor Anatolievich Zubow*

Chapter 5 Class Influence of Relative Content of Monterpenes
and Sesquiterpenes and its Impact on the Antioxidant
Activity: Applied for Some Algerian Endemic and
Medicinal Plants 77
Nadhir Gourine

Chapter 6	Chemical Composition and Biological Activities of Essential Oils from Three Medicinal Plants of Nigeria *Oladipupo A. Lawal, Isiaka A. Ogunwande, Bolanle T. Jinadu, Omolara T. Bejide and Emmanuel E. Essien*	95
Chapter 7	Fluorescence and Phosphorescence Spectra of Xanthone in the Vapor Phase *Takao Itoh*	121
Chapter 8	Mass Transfer Characteristics of an External Loop Airlift Reactor *Makaira Govender, Christopher Perumal, Elley M. Obwaka and Amir H Mohammadi*	131
Chapter 9	Applications of Isolation and Structure Elucidation of Secondary Metabolites in Natural Product Chemistry Laboratory *Aliefman Hakim and A. Wahab Jufri*	171
Index		**183**

PREFACE

The authors' of this latest volume discuss recent advances in chemistry research. Chapter One provides recent advances and developments in gliadin-based drug delivery systems. Chapter Two discusses advances achieved in the development of catalytic processes for the production of ß-citronellol derivatives as alkyl esters, aldehydes, carbamates, and carbonates. Chapter Three focuses on the developments and applications of a new type of polymer-supported organosilica layered-hybrid membranes. Chapter Four investigates the long-range order in natural oils and fats. Chapter Five reviews the possible link between the main classes' content of monoterpenes and sesquiterpenes and finds a direct link between them and their antioxidant activity. Chapter Six reports on the chemical compositions and biological activities of essential oils hydrodistilled from three plants from Nigeria. Chapter Seven reviews fluorescence and phosphorescence spectra of xanthone in the vapor phase. Chapter Eight examiners mass transfer characteristics of an external loop airlift reactor. Chapter Nine studies applications of isolation and structure elucidation of secondary metabolites in a natural product chemistry laboratory.

Chapter 1 - In the past few decades, pharmaceutical industry has explored protein-based nanoparticles as carriers in drug delivery and controlled release applications. Proteins have been used in drug release systems due to their exceptional characteristics, such as biodegradability, renewable sources, and binding capacity to several drugs. In recent years, gliadin stood out as a promising protein for the synthesis of nanoparticles and nanocapsules for drug delivery. Moreover, gliadins are hydrophobic proteins able to interact with skin keratin. One of the grounds is its low solubility in water (gliadin is soluble in alcohol/water mixtures). Besides, gliadin can adhere in the mucus layer of the stomach due to its strong adhesive capacity. However, the use of these macromolecules is sometimes limited because they are known to cause celiac disease in sensitive individuals. The mechanism of this disease, involves mucosal damage in the intestinal epithelium by both direct toxic action and immunological mechanism. In this chapter, the authors discussed the chemical structure of gliadin followed by a review of the main technological approaches and advances using gliadin as carrier in drug delivery systems.

The implications of using this protein in health, the main agent of celiac disease, will also be discussed.

Chapter 2 - The conversion of the natural origin alcohols to more value-added products has been a reaction of great interest for synthetic chemistry at academic and industrial level. In special, β-citronellol derivatives are widely used for preparation of pharmaceutical, insecticides, fragrance, and many others chemicals. β-citronellol is a common component of essential oils of citronella and rose, it emerges as a highly attractive raw material, due to its great abundance and affordability. Several chemicals of interest for plentiful industries can be obtained through different catalytic reactions with β-citronellol, such as alkyl esters or ethers, aldehydes, and epoxide-derivatives. Particularly, β-citronellyl esters are useful as fragrances and perfumes, while β-citronellyl carbamates are useful to synthesize insecticides. Herein, the authors have describing metal-catalyzed reactions that are a good alternative to the enzymatic processes, which are expensive and sensitive to the reactions conditions such as pH variations, solvent, and temperature. In this work, the authors wish the recent advances achieved in the development of catalytic processes for the production of β-citronellol derivatives as alkyl esters, aldehydes, carbamates, and carbonates. The authors will pay special attention to the development of homogeneous metal catalysts what are active under mild reaction conditions. Numerous industries in all parts of the world have crescent demand by developing of environmentally friendly technologies based on renewable raw materials, which becomes especially attractive when converted to valuable chemicals in the presence of catalysts active. The metal catalysts performance was assessed on the esterification, oxidation and urea alcoholysis reactions with β-citronellol. The development of innovative catalytic processes that may lead to high value-added products, based on renewable feedstock such as β-citronellol, is still a challenge to be overcome. The authors hope that this work can significantly contribute to improve this important research field.

Chapter 3 - A new type of polymer-supported organosilica layered-hybrid membrane was developed. Using 1,2-bis(triethoxysilyl)ethane (BTESE) as a single precursor, a uniform, thin and perm-selective organically bridged silica active layer was successfully deposited onto a porous polysulfone support via a facile and reliable sol–gel process. These new types of organosilica layered-hybrid membranes were then used for the vapor permeation (VP) dehydration of isopropanol-water (90/10 wt%) solutions, and showed a stable water flux of 2.3 kg/(m² h) and an improved separation factor of about 2500. Moreover, these layered-hybrid membranes also displayed good stability and reproducibility in the reverse osmosis (RO) desalination of a 2000 ppm NaCl solution process, and showed a stable and high degree of water permeability (approximately 1.2×10^{-12} m³ m⁻² s⁻¹ Pa⁻¹) with salt rejection that was competitive (96%) with conventional processing. It shows that the separation performances of these polymer-supported organosilica layered-hybrid membranes have been equal to, or even better than, many ceramic-supported membranes.

Chapter 4 - The long-range order (LRO) in vegetable oils (olive, sunflower, maize) and in some perfume oils (rose, lavender, vermouth, wormwood and lemongrass) was investigated using the gravitational mass spectroscopy (GMS). The LRO of them was compared with that one of butter and pig fat. In all samples, a hierarchy in LRO at the level of clusters and super cluster structures (sub micelles and micelles) prevailed, they were allowed by the gravitational noises (GN) of the universe and built to minimize the potential energy of the adhesive-cohesive interaction between molecules. The cluster formation was satisfactorily described by the first Zubow equation. For the mass range up to 200 million Dalton, the main LRO parameters (average molecular mass, number of cluster kinds, part of collapsed and expanded forms, cluster distribution) were analyzed.

Chapter 5 - Since the middle ages, essential oils have been widely used for bactericidal, virucidal, fungicidal, antiparasitical, insecticidal, medicinal and cosmetic applications, especially nowadays in pharmaceutical, sanitary, cosmetic, agricultural and food industries. Because of the mode of extraction, mostly by distillation from aromatic plants, they contain a variety of volatile molecules such as terpenes and terpenoids, phenol-derived aromatic components and aliphatic components. Most of the commercialized essential oils are chemotyped by gas chromatography and mass spectrometry analysis. Despite their wide use and being familiar to us as fragrances, it is important to develop a better understanding of the influence of the two main classes of mono- and sesqui-terpenes on the antioxidant activity of their essential oils. These finding if exists, could certainly be intended for exploring new ways of applications for food additives seeking a better human health. The aim of current chapter is to roughly investigate the possible link between the main classes' content of monoterpenes and sesquiterpenes and to try to find a direct link between them and their antioxidant activity. This investigation was carried out using the multivariate statistical analysis tool, and it was applied for a random selection of essential oils of some common medicinal and endemic plants growing in Algeria. The main important results were: first, there is an opposite linear correlation between the percentages of oxygenated monoterpenes and those of the monoterpenes hydrocarbons. Second, and for a unique case study of *Pistacia altantica* essential oils, the antioxidant activity power measured by the scavenging of free radicals of DPPH was in good correlation with oxygenated monoterpenes percentages.

Chapter 6 - The chemical compositions and biological activities of essential oils hydrodistilled from three plants from Nigeria are being reported. The constituents of the essential oils were analysed by means of gas chromatography (GC) and gas chromatography-mass spectrometry (GC-MS) techniques. The antibacterial, insecticidal, larvicidal and phytotoxicity studies were evaluated by standard procedures. The major components of the leaf oil of *Blighia sapida* K.D. Koenig were 6,10,14-trimethyl-2-pentadecanone (12.8%), geranyl acetone (12.0%), phytol (10.8%) and α-ionone (6.1%). The main constituents of the flower oil of *Thevetia peruviana* Pers (Apocynaceae) were β-ionone (44.5%), terpinen-4-ol (8.9%) and, terpinolene (8.3%). However, limonene (30.0%), δ-cadinene (13.3%), α-copaene (10.1%) and terpinolene (10.0%) were the main compounds of the leaf oil. Moreover, 1,8-cineole (54.1%) and α-terpineol (15.6%) represent the main compounds of the seeds of *Aframomum longiscapum* (Hook.f.) K. Schum while the pod consists mainly of linalool (34.3%), 1,8-cineole (15.7%) and β-pinene (11.0%). In addition, β-pinene (27.9%), β-caryophyllene (18.8%), caryophyllene oxide (12.2%) and α-pinene (9.7%) were the constituents occurring in higher proportions in the leaf.

The *B. sapida* oil exhibited strongest antibacterial activity against *Staphylococcus aurues* and *Escherichia coli* with the minimum inhibitory concentration (MIC) of 0.3 mg/mL, *Bacillus subtilis* (MIC, 0.6 mg/mL), *Pseudomonas* spp. (MIC, 1.2 mg/mL) and *Kliebsiella* spp. (MIC, 1.3 mg/mL). The studied essential oil of *A. longisparcum* oil displayed strong antibacterial activity against *B. subtilis* (MIC: leaf and pod, 0.6 mg/mL; seed, 0.3 mg/mL), *S. aurues* (MIC: leaf, 0.6 mg/mL; seed, 0.1 mg/mL; pod, 0.3 mg/mL), *E. coli* (MIC: leaf, 1.2 mg/mL; seed and pod, 0.3 mg/mL), *Kliebsiella* spp. (MIC: seed, 0.6 mg/mL; pod, 1.2 mg/mL), *Pseudomonas* spp. (MIC: leaf, seed and pod, 1.2 mg/mL) and *Proteus* spp. (MIC: seed, 1.2 mg/mL).

The essential oils of *B. sapida* and *T. peruviana* displayed strong insecticidal activity against *Sitophilus zeamais*. The lethal concentrations (LC$_{50}$) were 6.28 mg/L air (*B. sapida* leaf), 1.52 mg/L (*T. peruviana* flower) and 4.35 mg/L air (*T. peruviana* leaf). The result of larvicidal activity of *B. sapida* leaf oil against the fourth-in-star larvae of *Anopheles gambiae* revealed a lethal concentration (LC$_{50}$) of 11.61 mg/L air. The phytotoxicity activity *T. peruviana* essential oils on seeds germination and seedling growth of *Zea mays* indicated the oils to have percentage germination ranging from 76.6% to 100% against the seeds of *Zea mays* at 25 to 500 mg/mL in dose dependent manner.

Chapter 7 - Spectral measurements of molecules in the vapor phase provide intrinsic properties of molecules free from the influence of environment. Emission spectra of xanthone vapor measured at different temperatures are shown along with the excitation and absorption spectra. The emission consists of fluorescence from S_1 (n, π^*), $^1A_2$ and phosphorescence from T_1 (n, π^*), $^3A_2$ state overlapping each other. The overlapping spectrum was separated to extract only each of the fluorescence and phosphorescence spectra. The vibrational structures of both the fluorescence and phosphorescence were interpreted in terms of the C = O stretching mode and the modes combined with it. The S_1 and T_1 origins are located at 26940 and 25700 cm^{-1}, respectively. Analysis of the data includes the determination of the vibrational frequencies in the fluorescence and phosphorescence spectra.

Chapter 8 - Airlift reactors facilitate contact of a liquid with a gas or solid phase while providing good agitation, mass and heat transfer. They are applied in both chemical and biochemical industries. The aim of this study was to determine and compare the overall volumetric mass transfer characteristics for an external loop airlift reactor for four configurations, taking into account hydrodynamic properties such as gas hold up, superficial gas velocity, superficial liquid velocity and pressure drops. A total of twenty runs were conducted, five runs per configuration, with varying gas flowrate for each run from 0.5l/s to 2l/s. Flow rates above these values caused spillage from the disengagement tank. For each run, manometer readings and liquid flow rates were recorded in addition to dissolved oxygen concentrations obtained using a dissolved oxygen probe and YSI Data Manager Software. Thereafter, gas hold up, superficial gas and liquid velocity and overall volumetric mass transfer coefficient were calculated. The results acquired showed distinct trends. Increasing the superficial gas velocity increased the gas hold up, liquid circulation velocity and overall volumetric mass transfer coefficient in a linear manner. Configuration 1, only the riser, displayed the best results with a gas hold up of 0.8701 and overall volumetric mass transfer coefficient of 0.0420s^{-1}. Downcomer 1 in configuration 3 had the highest liquid circulation velocity of 1.41m/s as it had the smallest diameter. In general, configuration 1 showed the best results which can be attributed to the sparger being placed at the base of the riser. This allowed for direct contact of the gas with no bubbles disengaging in the disengagement tank, however, the downcomers are useful for a greater degree of mixing which was not achieved in the riser. Results are in accordance with literature trends. Outliers in data points were caused by deviations in flow rates used and a change in flow regime over a superficial gas velocity of 0.06m/s.

Chapter 9 - Natural products chemistry examines secondary metabolites contained in an organism, so it is strongly associated with pharmaceuticals, cosmetics, and pesticides. The chemical study of natural product based on experimental development and applications demanding high standards of laboratory activities. The laboratory activities involve the isolation of secondary metabolites from plants. The same secondary metabolites from a plant species can be isolated in a various ways, so there is no standard procedure to isolate the secondary metabolites of a plant species. These conditions can be used to train high-level thinking skills of learners. In natural product chemitry laboratory, learners can be given responsibility to undertake project to isolates the secondary metabolites from a variety of plant species. This laboratory works provide opportunities for students to design their own activities to isolate the secondary metabolites from medicinal plants. Students are exposed to skills as extraction, fractionation, purification, and structural elucidation of secondary metabolites. These laboratory activities can be useful for students at the third-year undergraduate level from many different disciplines including chemistry education, chemistry, pharmacy, and medicine.

In: Advances in Chemistry Research. Volume 35   ISBN: 978-1-53610-734-0
Editor: James C. Taylor   © 2017 Nova Science Publishers, Inc.

Chapter 1

RECENT ADVANCES AND DEVELOPMENT OF GLIADIN-BASED DRUG DELIVERY SYSTEM

Andresa da Costa Ribeiro*, Daiani Canabarro Leite, Rosane Michele Duarte Soares and Nádya Pesce da Silveira
Institute of Chemistry, Federal University of Rio Grande do Sul, Porto Alegre (RS), Brazil

ABSTRACT

In the past few decades, pharmaceutical industry has explored protein-based nanoparticles as carriers in drug delivery and controlled release applications. Proteins have been used in drug release systems due to their exceptional characteristics, such as biodegradability, renewable sources, and binding capacity to several drugs. In recent years, gliadin stood out as a promising protein for the synthesis of nanoparticles and nanocapsules for drug delivery. Moreover, gliadins are hydrophobic proteins able to interact with skin keratin. One of the grounds is its low solubility in water (gliadin is soluble in alcohol/water mixtures). Besides, gliadin can adhere in the mucus layer of the stomach due to its strong adhesive capacity. However, the use of these macromolecules is sometimes limited because they are known to cause celiac disease in

---

* Corresponding author: Phone: + 55 51 3308 9780, Phone: + 55 55 9119 8216, Fax: + 55 51 3308 7304, Email: desa.ribeiro@hotmail.com (Andresa da Costa Ribeiro).

sensitive individuals. The mechanism of this disease, involves mucosal damage in the intestinal epithelium by both direct toxic action and immunological mechanism. In this chapter, we discussed the chemical structure of gliadin followed by a review of the main technological approaches and advances using gliadin as carrier in drug delivery systems.

The implications of using this protein in health, the main agent of celiac disease, will also be discussed.

Keywords: gliadin, nanoparticles, celiac disease, controlled release

INTRODUCTION

In recent years, there has been a considerable interest in the development of novel drug delivery systems using nanotechnology (Elzoghby, Samy and Elgindy, 2012b). Different types of nano-sized carriers were developed for various drug-delivery applications (Elzoghby, Samy and Elgindy, 2012a). Among existing materials as nano-sized carriers stands out the natural polymers such as starch(Elvira, Mano, San Román and Reis, 2002), gelatin(Khan, Shukla and Bajpai, 2016), chitosan (Agnihotri, Mallikarjuna and Aminabhavi, 2004), zein (Joye, Davidov-Pardo, Ludescher and McClements, 2015), albumin (Elzoghby et al., 2012a), gliadin (Elzoghby et al., 2012b; Herrera, Veuthey and Dodero, 2016), etc. In general, nanocarriers may protect a drug from degradation, enhance the drug absorption by facilitating diffusion through epithelium, modify pharmacokinetic and drug tissue distribution profile and/or improve intracellular penetration and distribution (Elzoghby et al., 2012a). Furthermore, by modulating the surface properties, composition, the desired release pattern of the drug and its biodistribution can be achieved (Elzoghby et al., 2012a).

A targeted drug delivery for brain disease, for example, was developed by Kim and coworkers (Kim, Choi, Kim and Tae, 2013) using a chitosan-conjugated Pluronic-based nano-carrier with a specific target peptide for the brain. Using starch, El-Feky (GS, MH, MA, ME and A, 2015) et al. synthesized crosslinked starch nanoparticles as a carrier for indomethacin and acyclovir drugs. In a different research, a starch functionalized graphene pH-sensitive nano-carrier for drug delivery was produced (K. Liu, Wang, Li and Duan, 2015). Although the targeting strategies do not specify the size of drug carriers, it is generally known that nano-sized carriers are a prerequisite for efficient drug targeting (Yokoyama, 2005).

Several types of proteins have also been studied for their ability to produce nanoparticles (Joye, Nelis and McClements, 2015a), such as human serum albumin (Peng et al., 2016), gelatin (Khan et al., 2016), and zein (Fereshteh, Fathi, Bagri and Boccaccini, 2016). Proteins are interesting due to its large bioavailability, better encapsulation, and, controlled release (Kumari, Yadav and Yadav, 2010). Moreover, protein nanoparticles can be easily prepared and scaled up during manufacture (Elzoghby et al., 2012b). Gelatin nanoparticles, for example, have been reported as a good delivery of various drugs, such as anticancer (Muvaffak, Gurhan, Gunduz and Hasirci, 2005) and anti-inflammatory (Thakur et al., 2011).

Due to their multiple functional groups in the primary sequences of polypeptides, protein nanoparticles can be explored to create different interactions with therapeutic compounds and subsequently form three-dimensional networks offering a variety of possibilities for reversible binding of active molecules, protecting them in a matrix as well as targeting to the action site (Elzoghby et al., 2012b).

The interest in gliadin for biomedical applications, has increased, and since 1995(Stella, Vallée, Albrecht and Postaire, 1995) it has been used to produce nanoparticles(Miguel Angel Arangoa, Campanero, Renedo, Ponchel and Irache, 2001; Cécile Duclairoir et al., 1999; C. Duclairoir, Orecchioni, Depraetere, Osterstock and Nakache, 2003; Gulfam et al., 2012). This protein is able to adhere in the mucus layer of the stomach due to its strong adhesive capacity. Besides that, gliadin systems are rated as potential drug carriers in topical formulations (Teglia & Secchi, 1994), since they interact with epidermal keratin (Elzoghby et al., 2012b).

However, gliadin causes an autoimmune disorder named as Celiac disease. This disorder damages the inner surface of the small intestine and interfere in the nutrient absorption (Lee, Alwahab and Moazzam, 2013). As a segment of the population has wheat allergy the elimination of this protein from the daily diet is a potential drawback of using this protein (Joye, Nelis, et al., 2015a).

In this chapter we discuss the chemical structure and conformation of gliadin followed by a review of the main technological approaches and advances using gliadin as carrier in drug delivery systems.

CHEMICAL STRUCTURE OF THE GLIADINS AND THEIR CONFORMATION

Gliadin is a complex protein mixture present in wheat, rye and barley which is not fully degraded by humans (Herrera et al., 2016) and along with glutenin are gluten proteins (Rahaman, Vasiljevic and Ramchandran, 2016).

Gliadins comprise about half of the total prolamins of gluten, are monomeric proteins with intramolecular disulphide bonds and contribute to the viscous nature of doughs (Ang et al., 2010). Glutenin is a polymeric protein which are linked by inter-chain disulphide bonds (Rahaman et al., 2016) and contribute to dough cohesiveness and elasticity (Wang et al., 2016). The gliadins have a molecular weight of 30–60 kDa, and recent reports have shown that this protein is soluble in alcohol solutions (e.g., 60–70% ethanol (Vo Hong et al., 2016)) and its solubility in water is low (Herrera et al., 2016; Vo Hong et al., 2016).

According to the literature, different separation techniques have revealed that gliadin is a complex mixture of numerous protein components (H. Wieser, 1996). First attempts to separate this protein were made by starch gel electrophoresis or cation exchange chromatography. The four subfractions obtained were called α-, β-, γ- and ω-Gliadin (Kontogiorgos, 2011; Qi, Wei, Yue, Yan and Zheng, 2006; Sjöström, Friis, Norén and Anthonsen, 1992; Tatham & Shewry, 1985; H. Wieser, 1996; Herbert Wieser, 2007), in accordance with their electrophoretic mobility at acid pH (Ang et al., 2010; H. Liu et al., 2016). Comparisons of amino acid and DNA sequences show that the α- and β-gliadins are closely related and referred to as "α -type" gliadins (H. Wieser, 1996), while the γ - and ω-gliadins are structurally distinct (Ang et al., 2010). However, some studies suggested that gliadin is divided into α/β, γ and ω-Gliadin (Ferranti, Mamone, Picariello and Addeo, 2007; Popineau & Pineau, 1985; Qi et al., 2006). α/β and γ-gliadin are low molecular weight proteins (MW 28-35 kDa) with six and eight cysteine residues, respectively, whereas ω-Gliadin (MW 40-75 kDa) does not have cysteine (Rahaman et al., 2016). A-gliadins, a special group of α-gliadins, are monomeric at low pH and ionic strengths but at higher pH values the proteins self-assemble to form a fibrillary network stabilized by physical interactions (Kontogiorgos, 2011)

These proteins consist almost entirely of repetitive sequences rich in glutamine and proline (e.g., PQQPFPQQ). α/β- and γ-gliadins have overlapping molecular weight (≈ 28 000 – 35 000) and the proportions of glutamine and proline are much lower than those of ω –gliadins (Herbert Wieser, 2007).

The typical unit of ω-gliadins is QPQQPFP, which is repeated up to 16 times and interspersed by additional residues (Herbert Wieser, 2007). ω-gliadins are characterized by the presence of large proportions of glutamine, proline and phenylalanine residues (which together account for about 80% of the total residues) (Ferranti et al., 2007; Howdle, 2006; Herbert Wieser, 2007) but few or no residues of sulphur containing amino acids (cysteine and methionine). In contrast, α-, β-and γ-gliadins have less proline, glutamine and phenylalanine, but 2-3 mol. % cysteine plus methionine. Most, probably all, of the cysteine residues are involved in intramolecular disulphide bonds. Reduction of these residues causes changes in mobility at low pH (presumably due to changes in the conformation compactation), and it is not possible to recognize discrete groups of α-, β- and γ-gliadins (Peter R. Shewry, Tatham, Forde, Kreis and Miflin, 1986).

Each type of gliadin differs significantly in the contents of a few amino acids (each type has two different N and C-terminal domains). The N-terminal domain (40 – 50% of total proteins) consists mostly of repetitive sequences rich in glutamine, proline, phenylalanine and tyrosine and is unique for each type (Herbert Wieser, 2007). Although the distribution of total gliadins among the different types is strongly dependent on wheat variety (genotype) and growing conditions (soil, climate, fertilization), it can be generalized that α/β- and γ-gliadins are major components, whereas the ω-gliadins occur in much lower proportions(Herbert Wieser, 2007).

The repetitive units of α/β-gliadins is formed by three different regions: a short non-repetitive N-terminal domain, a central domain without cysteine, formed by repetition of two Pro- and Gln-rich sequences (heptapeptide: Pro-Gln-Pro-Gln-Pro-Phe-Pro and pentapeptide: Pro-Gln-Gln-Pro-Tyr) and finally a long C-terminal domain containing six Cys and most of the charged amino acid residues (Ferranti et al., 2007). Furthermore, studies on the secondary structure have indicated that the N-terminal domains of α/β- and ω-gliadins are characterized by β-turn conformation. The non-repetitive C-terminal domain contains considerable proportions of α-helix and β-sheet structures (Herbert Wieser, 2007). γ-gliadin are formed by three different regions: a short N-terminal domain, a repetitive central domain formed by repetition of a Pro- and Gln-rich sequence (heptapeptide Pro-Gln-Gln-Pro-Phe-Pro-Gln) and a C-terminal domain containing the eight Cys residues and most of the charged residues(Ferranti et al., 2007).

On the other hand, two-dimensional electrophoresis or reversed-phase high performance liquid chromatography allow the separation the gliadin into more than 100 components. Thus, they can be classified by their primary structures (complete or partial amino acid sequences, molecular masses, amino acid compositions) into four different protein type: ω5-, ω1,2-, α and γ-gliadin(Qi et al., 2006; H. Wieser, 1996). However, within each type, structural differences are small due to substitution, deletion and insertion of single amino acid residues (Herbert Wieser, 2007). The ω5- and ω1,2-type as well as the α- and γ-type are related, where ω5 have higher molecular masses and higher contents of glutamine and phenylalanine(H. Wieser, 1996) than the others (ω5-gliadin have higher molecular weight (≈ 50 000) than ω1,2-gliadin (≈ 40 000))(Herbert Wieser, 2007).

In other classification, Bietz et al.(Bietz, Huebner, Sanderson and Wall, 1977) and Kasarda et al.(Kasarda, Autran, Lew, Nimmo and Shewry, 1983) reclassified the gliadin into two groups: the S-poor (ω-gliadin) (Festenstein, Hay, Miflin and Shewry, 1985) and the S-rich (α-, β- (or α/β-) and γ-gliadins) (Kontogiorgos, 2011; Qi et al., 2006; Tatham & Shewry, 1985). The repeated sequences present in all wheat prolamins are largely responsible for the unusual amino acid compositions (being rich in proline and glutamine) and solubility properties of the whole proteins. Biophysical studies show that these repetitive domains are not globular (unlike the non-repetitive domains of the S-rich), but form unusual spiral super secondary structures based on repeated β-turns and (in the S-rich and S-poor types only) poly-L-proline II structure (P. R. Shewry, Tatham, Barcelo and Lazzeri). The low water solubility has been attributed to the presence of disulphide bonds and to the cooperative hydrophobic interactions which cause the protein chains to assume a folded shape (Elzoghby et al., 2012b).

In conclusion, more and more gliadins have been characterized, and the gliadin family is apparently much larger and more diverse than previously thought. Classification of gliadins is very complex, and further investigations are necessary for better understanding (Qi et al., 2006).

TECHNOLOGICAL APPROACHES AND ADVANCES USING GLIADIN AS CARRIER IN DRUG DELIVERY SYSTEMS

Drug delivery systems are interesting formulations to prevent numerous drawbacks related to the drug itself. In such a dosage form, the drug is gradually released unlike all conventional formulations (C. Duclairoir et al., 2003). Innovative technology for carriers demands the development of sophisticated drug delivery systems, which in turn need functional excipients that are able to produce delivery systems with specific drug release patterns (Scholtz, Van der Colff, Steenekamp, Stieger and Hamman, 2014).

Gliadin seems to be a promise macromolecule in the development of formulations with drugs. Its hydrophobicity and solubility in certain conditions permits the design of nanoparticles capable of protecting the loaded drug and controlling its release (Jahanshahi & Babaei, 2008). Furthermore, as a biopolymer, it does not present the common drawback of synthetic materials. In addition, they are able to interact with skin keratin (C. Duclairoir et al., 2003).

The ability to solubilize gliadin in a certain pH allows to be used in site-specific drug delivery system, for example. Targeting the release of a drug to a specific site in the gastrointestinal tract can greatly improve the therapeutic efficacy and reduce the side effects of certain drugs, as reported in the literature with others systems (He, Du, Cao, Xiang and Fan, 2008; Smrdel, Cerne, Bogataj, Urleb and Mrhar, 2010). Other advantages of the use of gliadin (and proteins, in general) in drug delivery systems the possibility to carry different charges on its surface (Narendra Reddy & Yang, 2011). For example, nanoparticles containing proteins on their surface are targeted to cancer cells because proteins can bind to cancer cell surface receptor proteins (Davis, Chen and Shin, 2008).

In the last 20 years, the development and characterization of delivery systems with gliadin have been explored and a summary of these studies are described on Table 1.

In these studies, guest molecules such as all-trans-retinoic acid (Ezpeleta et al., 1996), clarithromycin, omeprazole (Sharma et al., 2014), paclitaxel, lysozyme (Fajardo et al., 2014; Stella et al., 1995), vitamin E, linallol/linalyl acetate mixture, and benzakonium (C. Duclairoir et al., 2003), among others, have been employed as drugs using gliadin as carrier.

The first paper on the topic was published in 1995(Stella et al., 1995). In this study, soft capsules and chewable gums of gliadin were prepared and used to encapsulate paracetamol. The study showed that gliadin appears to be a highly promising carrier with low cost and bio acceptable for the manufacture of drug formulations with a very interesting controlled released potency.

Table 1. Gliadin as drug delivery system

Year	Article
1995	Gliadin films. I. Preparation and *in vitro* evaluation as a carrier for controlled drug release (Stella et al., 1995)
1996	Gliadin nanoparticles for the controlled release of all-*trans*-retinoic acid (Ezpeleta et al., 1996)
1998	Formation of gliadin nanoparticles: influence of the solubility parameter of the protein solvent (C. Duclairoir, Nakache, Marchais and Orecchioni, 1998)
1999	Preparation of *Ulex europaeus* lectin-gliadin nanoparticle conjugates and their interaction with gastrointestinal mucus (Ezpeleta et al., 1999)
1999	Electrophoretic separation and characterization of gliadin fractions from isolates and nanoparticulate drug delivery systems (M. A. Arangoa, Campanero, Popineau and Irache, 1999)
2000	Bioadhesive potential of gliadin nanoparticulate systems (M. A. Arangoa et al., 2000)
2001	Gliadin nanoparticles as carriers for the oral administration of lipophilic drugs. Relationships between bioadhesion and pharmacokinetics (Miguel Angel Arangoa et al., 2001)
2002	Gliadin matrices for microencapsulation processes by simple coacervation method (Mauguet et al., 2002)
2003	Evaluation of gliadin nanoparticles as drug delivery systems: a study of three different drugs (C. Duclairoir et al., 2003)
2003	Receptor mediated targeting of lectin conjugated gliadin nanoparticles in the treatment of *Helicobacter pylori* (Umamaheshwari & Jain, 2003)
2006	Clarithromycin based oral sustained release nanoparticulate drug delivery system (Ramteke, Maheshwari and Jain, 2006)
2008	Self-crosslinked gliadin fibers with high strength and water stability for potential medical applications (N. Reddy & Yang, 2008)
2008	Clarithromycin-and omeprazole-containing gliadin nanoparticles for the treatment of Helicobacter pylori (Ramteke & Jain, 2008)
2011	Preparation of tetanus toxoid and ovalbumin loaded gliadin nanoparticles for oral immunization (Kajal & Misra, 2011)
2012	Anticancer drug-loaded gliadin nanoparticles induce apoptosis in breast cancer cells (Gulfam et al., 2012)
2014	Preparation and characterization of Paclitaxel-loaded gliadin nanoparticles(Sharma, Deevenapalli, Singh, Chourasia and Bathula, 2014)
2014	Chemically modified gliadins as sustained release systems for lysozyme (Fajardo, Balaguer, Gomez-Estaca, Gavara and Hernandez-Munoz, 2014)
2015	Gliadin-based nanoparticles: Fabrication and stability of food-grade colloidal delivery systems (Joye, Nelis, et al., 2015a)
2015	Gliadin-based nanoparticles: stabilization of food-grade post-production polysaccharide coating (Joye, Nelis and McClements, 2015b)

In the next year, Ezpeleta (Ezpeleta et al., 1996) et al. developed gliadin nanoparticles to encapsulate all-trans-retinoic acid (RA). This study showed the preparation of gliadin particles through the coacervation method was able to produce spherical and submicron-sized particles. Thus, the study proved that gliadin is a delivery system for all-trans-retinoic acid.

In the range of 1998 - 2006(M. A. Arangoa et al., 1999; Miguel Angel Arangoa et al., 2001; M. A. Arangoa et al., 2000; C. Duclairoir et al., 1998; C. Duclairoir et al., 2003; Ezpeleta et al., 1999; Ezpeleta et al., 1996; Mauguet et al., 2002; Ramteke et al., 2006; Umamaheshwari & Jain, 2003) the desolvation method was used for gliadin nanoparticles development and for the controlled release systems for hydrophobic and amphiphilic drugs.

Mauguet et al. (Mauguet et al., 2002) used the coacervation method for gliadin nanoparticles preparation for hexadecane (oil) encapsulation. Hexadecane was emulsified by a gliadin solution and the coacervation phenomenon was induced adding a salt solution in the continuous phase of the emulsion containing gliadin. The main problem of the microencapsulation process by salting-out was to control the capsule size and the agglomeration of them. This study succeeded in preventing the agglomeration phenomenon by adjusting the kinetics of the salt addition. The effect of different process parameters (gliadin concentration, quantity and addition kinetics of the coacervation agent, and cross-linker concentration) was studied regarding the final microcapsule characteristics (shape, size, composition, and mechanical resistance evaluated by a centrifugation test). The systems showed suitable for delivery systems utilization.

Different methods were used to develop delivery systems with gliadin in the last 10 years (N. Reddy & Yang, 2008). For example, self-crosslinked fibers which showed mechanical properties similar to wool and better than mechanical properties of zein, soyprotein and collagen fibers. Moreover, gliadin fibers were conductive and able to promote cell attachment, growth, and proliferation. In 2012, Gulfam et al. (Gulfam et al., 2012) employed the electrospray deposition method. This process is similar to the electrospinning technique. However, electrospray deposition can be created when the viscosity of the liquid is sufficient low. The method showed that breast cancer cells cultured with cyclophosphamide-loaded 7% gliadin nanoparticles for 24 h became apoptotic. Therefore, the gliadin-based nanoparticle could be a powerful tool for delivery and controlled release of anticancer drugs.

In 2014 (Fajardo et al., 2014), a different method was used for buildup delivery systems with gliadin: gliadin films. Until that moment there were no previous reports on the use of gluten or gliadin films as carriers. Thus, in this study the objective was to characterize the release kinetics of lysozyme from gliadin films cross-linked with cinnamaldehyde. The effect of lysozyme incorporation on both the mechanical and barrier properties (water vapor and oxygen) of the films was also evaluated. As result, the gliadin films after incorporation preserved their integrity in water. The release rate of the antimicrobial agent was controlled by the reticulation of the protein matrix and films with a freely cross-linked structure releasing a greater amount of lysozyme, exhibiting high antimicrobial activity.

Finally, in 2015 (Joye, Nelis, et al., 2015a, 2015b), two studies were published by Joey et al. The scope of these researches was based in the gliadin nanoparticles preparation using a non-solvent precipitation which could be used for food products. Studies showed that the gliadin nanoparticles prepared by this method may be useful in development of delivery systems to encapsulate, protect, target and release active ingredients for use in the food industry.

Summing up, it is believed that the gliadin presents a great potential as biomaterial due to its properties, and can be used as carriers for many types of drugs. However, it is necessary more research concerning the structure and toxicity due to their implication in health as main agent of celiac disease.

IMPLICATIONS OF THE USE OF GLIADIN IN HEALTH

Celiac disease is a common chronic gastrointestinal disorder both in children and adults (Sjöström et al., 1992). Although the classical features of celiac disease have been described more than two centuries ago, only in 1950 it was identified that eating wheat products was harmful to celiac patients (H. Wieser, 1996). Samuel Gee was the first to clearly define celiac disease and to recognize the role of diet. In 1888 he stated, 'The allowance of farinaceous foods must be small, but if the patient can be cured at all, it must be by means of diet.' Sixty years later, the Dutch pediatrician Willem Karel Dicke suggested a direct role for gluten and described the histological intestinal alterations in celiac disease. In the 1980s, the evidence for a primary association of celiac disease with particular DQ molecules was described.

Over the past two decades, celiac disease has emerged as a major public health problem in many countries. In North America and Europe, the prevalence of celiac disease in the general population has been shown to be at-least 1% (La Vieille, Pulido, Abbott, Koerner and Godefroy, 2016; Singh, Arora, Singh, Strand and Makharia, 2016). Furthermore, there are intercountry differences in the prevalence of celiac disease in Europe ranging from 0.3% in Germany to 2% in Finland, and the reasons for the differences in the prevalence cannot be explained convincingly based on differences in the pattern of gluten intake and celiac disease predisposing human leukocyte antigen (HLA) haplotypes. Celiac disease has been reported to be common in the Pacific nations that are mostly populated by individuals of European origin (e.g., Australia and New Zealand). The prevalence of celiac disease is almost similar to that in other European countries even in Latin American countries, such as Brazil, having relatively higher proportion of Caucasian population(Marciano, Savoia and Vajro, 2016; Singh et al., 2016).

Studies have revealed that celiac disease occurs in adults and children at rates of 1%. It is, nevertheless, considered as one of most common chronic diseases. Celiac disease can be diagnosed at any age. Infants and young children present diarrhea, abdominal distension, and failure to thrive. However, vomiting, irritability, anorexia, and constipation are also common. Older children often present extra intestinal manifestations, such as short stature, neurologic symptoms, or anemia. The mean age of diagnosis in adults is between 40 and 50 years old. Women are diagnosed at two to three times the rate of men, though this gender predominance is lost in older individuals. The most frequent single mode of presentation in adults is diarrhea, though this mode of presentation accounts for less than 50% of cases (Jabri, Kasarda and Green, 2005).

An important result of these early studies was the recognition that celiac toxicity is closely related to the taxonomy of cereals (these studies showed that the gluten fraction of wheat was toxic, whereas the starch and albumin fractions were not (Howdle, 2006)). Moreover, the toxic factor of wheat was found in the gluten, in particular, in the gliadin fraction. This occurs because studies indicate that the toxic prolamins of wheat are characterized by high contents of glutamine ($\approx$ 36%) and proline (17 - 23%), and the gliadin is rich of these compounds. For these reasons, gliadin was the main subject of studies on the relation between protein structure and celiac toxicity (H. Wieser, 1996).

The symptoms of the disease are developed after the intake of foods containing gluten and it triggers an inflammatory state of the duodenal mucosa, which result is reduced intestinal villus height and hyperplastic cryptae that may lead to complete villus atrophy (Sjöström et al., 1992; Tonutti & Bizzaro, 2014). Until the present moment, there is no cure for celiac disease. Therefore, people who suffer from this disease need to maintain a strict gluten-free diet for their entire life (Jabri et al., 2005; van Eckert et al., 2006). Thus, the knowledge of the chemical structure of the complex gliadin fraction is of particular importance to improve our understanding at a molecular level of the events which trigger this disease (Ferranti et al., 2007).

REFERENCES

Agnihotri, S. A., Mallikarjuna, N. N. and Aminabhavi, T. M. (2004). Recent advances on chitosan-based micro- and nanoparticles in drug delivery. *Journal of Controlled Release*, *100*(1), 5-28.

Ang, S., Kogulanathan, J., Morris, G. A., Kok, M. S., Shewry, P. R., Tatham, A. S. and Harding, S. E. (2010). Structure and heterogeneity of gliadin: a hydrodynamic evaluation. *European Biophysics Journal with Biophysics Letters*, *39*(2), 255-261.

Arangoa, M. A., Campanero, M. A., Popineau, Y. and Irache, J. M. (1999). Electrophoretic separation and characterisation of gliadin fractions from isolates and nanoparticulate drug delivery systems. *Chromatographia*, *50*(3-4), 243-246.

Arangoa, M. A., Campanero, M. A., Renedo, M. J., Ponchel, G. and Irache, J. M. (2001). Gliadin nanoparticles as carriers for the oral administration of lipophilic drugs. Relationships between bioadhesion and pharmacokinetics. *Pharmaceutical research*, *18*(11), 1521-1527.

Arangoa, M. A., Ponchel, G., Orecchioni, A. M., Renedo, M. J., Duchêne, D. and Irache, J. M. (2000). Bioadhesive potential of gliadin nanoparticulate systems. *European Journal of Pharmaceutical Sciences*, *11*(4), 333-341.

Bietz, J. A., Huebner, F. R., Sanderson, J. E. and Wall, J. S. (1977). Wheat gliadin homology revealed through N-terminal amino acid sequence analysis. *Cereal Chemistry*.

Davis, M. E., Chen, Z. and Shin, D. M. (2008). Nanoparticle therapeutics: an emerging treatment modality for cancer. *Nat Rev Drug Discov*, *7*(9), 771-782.

Duclairoir, C., Irache, J. M., Nakache, E., Orecchioni, A. M., Chabenat, C. and Popineau, Y. (1999). Gliadin nanoparticles: formation, all-trans-retinoic acid entrapment and release, size optimization. *Polymer international*, *48*(4), 327-333.

Duclairoir, C., Nakache, E., Marchais, H. and Orecchioni, A. M. (1998). Formation of gliadin nanoparticles: influence of the solubility parameter of the protein solvent. *Colloid and Polymer Science*, *276*(4), 321-327.

Duclairoir, C., Orecchioni, A. M., Depraetere, P., Osterstock, F. and Nakache, E. (2003). Evaluation of gliadins nanoparticles as drug delivery systems: a study of three different drugs. *International journal of pharmaceutics, 253*(1), 133-144.

Elvira, C., Mano, J. F., San Román, J. and Reis, R. L. (2002). Starch-based biodegradable hydrogels with potential biomedical applications as drug delivery systems. *Biomaterials, 23*(9), 1955-1966.

Elzoghby, A. O., Samy, W. M. and Elgindy, N. A. (2012a). Albumin-based nanoparticles as potential controlled release drug delivery systems. *Journal of Controlled Release, 157*(2), 168-182.

Elzoghby, A. O., Samy, W. M. and Elgindy, N. A. (2012b). Protein-based nanocarriers as promising drug and gene delivery systems. *Journal of Controlled Release, 161*(1), 38-49.

Ezpeleta, I., Arangoa, M. A., Irache, J. M., Stainmesse, S., Chabenat, C., Popineau, Y. and Orecchioni, A. M. (1999). Preparation of Ulex europaeus lectin-gliadin nanoparticle conjugates and their interaction with gastrointestinal mucus. *International journal of pharmaceutics, 191*(1), 25-32.

Ezpeleta, I., Irache, J. M., Stainmesse, S., Chabenat, C., Gueguen, J., Popineau, Y. and Orecchioni, A. M. (1996). Gliadin nanoparticles for the controlled release of all-trans-retinoic acid. *International Journal of Pharmaceutics, 131*(2), 191-200.

Fajardo, P., Balaguer, M. P., Gomez-Estaca, J., Gavara, R. and Hernandez-Munoz, P. (2014). Chemically modified gliadins as sustained release systems for lysozyme. *Food Hydrocolloids, 41,* 53-59.

Fereshteh, Z., Fathi, M., Bagri, A. and Boccaccini, A. R. (2016). Preparation and characterization of aligned porous PCL/zein scaffolds as drug delivery systems via improved unidirectional freeze-drying method. *Materials Science and Engineering: C, 68,* 613-622.

Ferranti, P., Mamone, G., Picariello, G. and Addeo, F. (2007). Mass spectrometry analysis of gliadins in celiac disease. *Journal of mass spectrometry, 42*(12), 1531-1548.

Festenstein, G. N., Hay, F. C., Miflin, B. J. and Shewry, P. R. (1985). Specificity of an antibody to a subunit of high-molecular-weight storage protein from wheat seed and its reaction with other cereal storage proteins (prolamins). *Planta, 164*(1), 135-141.

GS, E. F., MH, E. R., MA, E. S., ME, E. N. and A, H. (2015). Utilization of Crosslinked Starch Nanoparticles as a Carrier for Indomethacin and Acyclovir Drugs. *Journal of Nanomedicine & Nanotechnology, 6*(1), 8.

Gulfam, M., Kim, J. E., Lee, J. M., Ku, B., Chung, B. H. and Chung, B. G. (2012). Anticancer Drug-Loaded Gliadin Nanoparticles Induce Apoptosis in Breast Cancer Cells. *Langmuir, 28*(21), 8216-8223.

He, W., Du, Q., Cao, D. Y., Xiang, B. and Fan, L. F. (2008). Study on colon-specific pectin/ethylcellulose film-coated 5-fluorouracil pellets in rats. *Int J Pharm, 348*(1-2), 35-45.

Herrera, M. G., Veuthey, T. V. and Dodero, V. I. (2016). Self-organization of gliadin in aqueous media under physiological digestive pHs. *Colloids and Surfaces B: Biointerfaces, 141*, 565-575.

Howdle, P. D. (2006). Gliadin, glutenin or both? The search for the Holy Grail in coeliac disease. *European journal of gastroenterology & hepatology, 18*(7), 703-706.

Jabri, B., Kasarda, D. D. and Green, P. H. R. (2005). Innate and adaptive immunity: the yin and yang of celiac disease. *Immunological reviews, 206*(1), 219-231.

Jahanshahi, M. and Babaei, Z. (2008). Protein nanoparticle: a unique system as drug delivery vehicles. *African Journal of Biotechnology, 7*(25).

Joye, I. J., Davidov-Pardo, G., Ludescher, R. D. and McClements, D. J. (2015). Fluorescence quenching study of resveratrol binding to zein and gliadin: towards a more rational approach to resveratrol encapsulation using water-insoluble proteins. *Food chemistry, 185*, 261-267.

Joye, I. J., Nelis, V. A. and McClements, D. J. (2015a). Gliadin-based nanoparticles: Fabrication and stability of food-grade colloidal delivery systems. *Food Hydrocolloids, 44*, 86-93.

Joye, I. J., Nelis, V. A. and McClements, D. J. (2015b). Gliadin-based nanoparticles: Stabilization by post-production polysaccharide coating. *Food Hydrocolloids, 43*, 236-242.

Kajal, H. and Misra, A. (2011). Preparation of tetanus toxoid and ovalbumin loaded gliadin nanoparticles for oral immunization. *Journal of biomedical nanotechnology, 7*(1), 211-212.

Kasarda, D. D., Autran, J. C., Lew, E. J. L., Nimmo, C. C. and Shewry, P. R. (1983). N-terminal amino acid sequences of ω-gliadins and ω-secalins: implications for the evolution of prolamin genes. *Biochimica et Biophysica Acta (BBA)-Protein Structure and Molecular Enzymology, 747*(1-2), 138-150.

Khan, H., Shukla, R. N. and Bajpai, A. K. (2016). Genipin-modified gelatin nanocarriers as swelling controlled drug delivery system for *in vitro* release of cytarabine. *Materials Science and Engineering: C, 61*, 457-465.

Kim, J. Y., Choi, W. I., Kim, Y. H. and Tae, G. (2013). Brain-targeted delivery of protein using chitosan- and RVG peptide-conjugated, pluronic-based nano-carrier. *Biomaterials*, *34*(4), 1170-1178.

Kontogiorgos, V. (2011). Microstructure of hydrated gluten network. *Food Research International*, *44*(9), 2582-2586.

Kumari, A., Yadav, S. K. and Yadav, S. C. (2010). Biodegradable polymeric nanoparticles based drug delivery systems. *Colloids and Surfaces B: Biointerfaces*, *75*(1), 1-18.

La Vieille, S., Pulido, O. M., Abbott, M., Koerner, T. B. and Godefroy, S. (2016). Celiac Disease and Gluten-Free Oats: A Canadian Position Based on a Literature Review. *Canadian Journal of Gastroenterology and Hepatology*, 2016.

Lee, S., Alwahab, N. S. A. and Moazzam, Z. M. (2013). Zein-based oral drug delivery system targeting activated macrophages. *International Journal of Pharmaceutics*, *454*(1), 388-393.

Liu, H., Wang, K., Xiao, L., Wang, S., Du, L., Cao, X. and Ye, X. (2016). Comprehensive Identification and Bread-Making Quality Evaluation of Common Wheat Somatic Variation Line AS208 on Glutenin Composition. *PloS one*, *11*(1), e0146933.

Liu, K., Wang, Y., Li, H. and Duan, Y. (2015). A facile one-pot synthesis of starch functionalized graphene as nano-carrier for pH sensitive and starch-mediated drug delivery. *Colloids Surf B Biointerfaces*, *128*, 86-93.

Marciano, F., Savoia, M. and Vajro, P. (2016). Celiac disease-related hepatic injury: Insights into associated conditions and underlying pathomechanisms. *Digestive and Liver Disease*, *48*(2), 112-119.

Mauguet, M. C., Legrand, J., Brujes, L., Carnelle, G., Larre, C. and Popineau, Y. (2002). Gliadin matrices for microencapsulation processes by simple coacervation method. *Journal of microencapsulation*, *19*(3), 377-384.

Muvaffak, A., Gurhan, I., Gunduz, U. and Hasirci, N. (2005). Preparation and characterization of a biodegradable drug targeting system for anticancer drug delivery: Microsphere-antibody conjugate. *Journal of drug targeting*, *13*(3), 151-159.

Peng, Z. Li. S., Han, X., Al-Youbi, A. O., Bashammakh, A. S., El-Shahawi, M. S. and Leblanc, R. M. (2016). Determination of the composition, encapsulation efficiency and loading capacity in protein drug delivery systems using circular dichroism spectroscopy. *Analytica Chimica Acta*, *937*, 113-118.

Popineau, Y. and Pineau, F. (1985). Fractionation and characterisation of γ-gliadins from bread wheat. *Journal of cereal science*, *3*(4), 363-378.
Qi, P. F., Wei, Y. M., Yue, Y. W., Yan, Z. H. and Zheng, Y. L. (2006). Biochemical and molecular characterization of gliadins. *Molecular Biology*, *40*(5), 713-723.
Rahaman, T., Vasiljevic, T. and Ramchandran, L. (2016). Shear, heat and pH induced conformational changes of wheat gluten – Impact on antigenicity. *Food Chemistry*, *196*, 180-188.
Ramteke, S. and Jain, N. K. (2008). Clarithromycin-and omeprazole-containing gliadin nanoparticles for the treatment of Helicobacter pylori. *Journal of drug targeting*, *16*(1), 65-72.
Ramteke, S., Maheshwari, R. B. U. and Jain, N. K. (2006). Clarithromycin based oral sustained release nanoparticulate drug delivery system. *Indian journal of pharmaceutical sciences*.
Reddy, N. and Yang, Y. (2011). Potential of plant proteins for medical applications. *Trends in Biotechnology*, *29*(10), 490-498.
Reddy, N. and Yang, Y. Y. (2008). Self-crosslinked gliadin fibers with high strength and water stability for potential medical applications. *Journal of Materials Science-Materials in Medicine*, *19*(5), 2055-2061.
Scholtz, J., Van der Colff, J., Steenekamp, J., Stieger, N. and Hamman, J. (2014). More Good News About Polymeric Plant-and Algae-Derived Biomaterials in Drug Delivery Systems. *Current Drug Targets*, *15*(5), 486-501.
Sharma, K., Deevenapalli, M., Singh, D., Chourasia, M. K. and Bathula, S. R. (2014). Preparation and characterization of paclitaxel-loaded gliadin nanoparticles. *Journal of Biomaterials and Tissue Engineering*, *4*(5), 399-404.
Shewry, P. R., Tatham, A. S., Barcelo, P. and Lazzeri, P. Molecular and cellular techniques in wheat improvement. (pp. 227-240).
Shewry, P. R., Tatham, A. S., Forde, J., Kreis, M. and Miflin, B. J. (1986). The classification and nomenclature of wheat gluten proteins: a reassessment. *Journal of Cereal Science*, *4*(2), 97-106.
Singh, P., Arora, S., Singh, A., Strand, T. A. and Makharia, G. K. (2016). Prevalence of celiac disease in Asia: A systematic review and meta-analysis. *Journal of gastroenterology and hepatology*, *31*(6), 1095-1101.
Sjöström, H., Friis, S. U., Norén, O. and Anthonsen, D. (1992). Purification and characterisation of antigenic gliadins in coeliac disease. *Clinica Chimica Acta*, *207*(3), 227-237.

Smrdel, P., Cerne, M., Bogataj, M., Urleb, U. and Mrhar, A. (2010). Enhanced therapeutic effect of LK-423 in treating experimentally induced colitis in rats when administered in colon delivery microcapsules. *J Microencapsul*, *27*(7), 572-582.

Stella, V., Vallée, P., Albrecht, P. and Postaire, E. (1995). Gliadin films. I: Preparation and *in vitro* evaluation as a carrier for controlled drug release. *International Journal of Pharmaceutics*, *121*(1), 117-121.

Tatham, A. S. and Shewry, P. R. (1985). The conformation of wheat gluten proteins. The secondary structures and thermal stabilities of α-, β-, γ- and ω-Gliadins. *Journal of Cereal Science*, *3*(2), 103-113.

Teglia, A. and Secchi, G. (1994). New protein ingredients for skin detergency: native wheat protein–surfactant complexes. *International Journal of Cosmetic Science*, *16*(6), 235-246.

Thakur, G., Mitra, A., Rousseau, D., Basak, A., Sarkar, S. and Pal, K. (2011). Crosslinking of gelatin-based drug carriers by genipin induces changes in drug kinetic profiles *in vitro*. *Journal of Materials Science: Materials in Medicine*, *22*(1), 115-123.

Tonutti, E. and Bizzaro, N. (2014). Diagnosis and classification of celiac disease and gluten sensitivity. *Autoimmunity Reviews*, *13*(4–5), 472-476.

Umamaheshwari, R. B. and Jain, N. K. (2003). Receptor mediated targeting of lectin conjugated gliadin nanoparticles in the treatment of Helicobacter pylori. *Journal of drug targeting*, *11*(7), 415-424.

van Eckert, R., Berghofer, E., Ciclitira, P. J., Chirdo, F., Denery-Papini, S., Ellis, H. J. and Wieser, H. (2006). Towards a new gliadin reference material–isolation and characterisation. *Journal of Cereal Science*, *43*(3), 331-341.

Vo Hong, N., Trujillo, E., Puttemans, F., Jansens, K. J. A., Goderis, B., Van Puyvelde, P. and Van Vuure, A. W. (2016). Developing rigid gliadin based biocomposites with high mechanical performance. *Composites Part A: Applied Science and Manufacturing*, *85*, 76-83.

Wang, K., Luo, S., Cai, J., Sun, Q., Zhao, Y., Zhong, X. and Zheng, Z. (2016). Effects of partial hydrolysis and subsequent cross-linking on wheat gluten physicochemical properties and structure. *Food Chemistry*, *197*, Part A, 168-174.

Wieser, H. (1996). Relation between gliadin structure and coeliac toxicity. *Acta Paediatrica*, *85*(s412), 3-9.

Wieser, H. (2007). Chemistry of gluten proteins. *Food Microbiology*, *24*(2), 115-119.

Yokoyama, M. (2005). Drug targeting with nano-sized carrier systems. *Journal of Artificial Organs*, *8*(2), 77-84.

In: Advances in Chemistry Research. Volume 35 ISBN: 978-1-53610-734-0
Editor: James C. Taylor © 2017 Nova Science Publishers, Inc.

Chapter 2

LEWIS ACID METAL SALTS-CATALYZED SYNTHESIS OF β-CITRONELLOL DERIVATIVES

Márcio José da Silva[*]
Universidade Federal de Viçosa, Chemistry Department,
Viçosa, Minas Gerais, Brazil

ABSTRACT

The conversion of the natural origin alcohols to more value-added products has been a reaction of great interest for synthetic chemistry at academic and industrial level. In special, β-citronellol derivatives are widely used for preparation of pharmaceutical, insecticides, fragrance, and many others chemicals. β-citronellol is a common component of essential oils of citronella and rose, it emerges as a highly attractive raw material, due to its great abundance and affordability. Several chemicals of interest for plentiful industries can be obtained through different catalytic reactions with β-citronellol, such as alkyl esters or ethers, aldehydes, and epoxide-derivatives. Particularly, β-citronellyl esters are useful as fragrances and perfumes, while β-citronellyl carbamates are useful to synthesize insecticides. Herein, we have describing metal-catalyzed reactions that are a good alternative to the enzymatic processes,

---

[*] silvamj2003@ufv.br.

which are expensive and sensitive to the reactions conditions such as pH variations, solvent, and temperature. In this work, we wish the recent advances achieved in the development of catalytic processes for the production of β-citronellol derivatives as alkyl esters, aldehydes, carbamates, and carbonates. We will pay special attention to the development of homogeneous metal catalysts what are active under mild reaction conditions. Numerous industries in all parts of the world have crescent demand by developing of environmentally friendly technologies based on renewable raw materials, which becomes especially attractive when converted to valuable chemicals in the presence of catalysts active. The metal catalysts performance was assessed on the esterification, oxidation and urea alcoholysis reactions with β-citronellol. The development of innovative catalytic processes that may lead to high value-added products, based on renewable feedstock such as β-citronellol, is still a challenge to be overcome. The authors hope that this work can significantly contribute to improve this important research field.

Keywords: β-citronellyl esters, tin(II) catalysts, palladium(II) catalysts, homogeneous catalysis

1. GENERAL INTRODUCTION

1.1. Monoterpene Alcohols

Monoterpenes are renewable compounds abundant on nature, being thus an affordable and inexpensive raw material. They usually have pleasant aroma, and are present in almost plants, mainly in the essential oils [1]. In addition, the cellulose industry generates wastes that are essentially constituted for α and β-pinene, which are monoterpenes widely useful as building blocks in fragrances and fine chemical industries [2-5].

On this sense, the conversion of biomass derivatives to chemicals and renewable fuels became an attractive alternative to the use of petrol derivatives [4, 5]. Nonetheless, it always desirable that the biomass conversion to chemicals occurrs through selective and green catalytic processes [6, 7].

Oxygenates monoterpenes derivatives (i.e., namely "monoterpenoids") are important for the production of pharmaceuticals, pesticides and biologically active compounds [8-11]. In particular, β-citronellol is a starting material for fragrances, agrochemicals and pharmaceuticals industries [12, 13]. This acyclic terpenic alcohol has been extensively used as a valuable feedstock for the perfume, beverage, food, and pharmaceutical industries.

β-Citronellyl acetate is an important ingredient for flavor and fragrances and has been industrially produced through enzyme-catalyzed processes [14, 15]. The immobilization of lipases has pivotal importance in the development of enzymatic reactions [16]. Lipases are selective catalysts, however, requires an adequate choice of solvent, besides rigid control of the temperature and pH of reaction medium [17-19]. Moreover, enzymatic catalysts are expensive, and the impossibility of catalyst recovery difficult its application in a large scale.

The synthesis of β-citronellyl acetate could be even more attractive when carried out through environmentally benign route, under solvent-free conditions and where the steps to purify products and recover the catalyst are simplified, minimizing the generation of residues and effluents [20, 21].

Lewis acid metals are always attractive alternative catalysts; they are less corrosive than Brønsted acid and more stable than enzymatic catalysts. Lewis acids are commercially available, in addition to be easily supported on solid matrixes [22]. On this regard, it is noteworthy that to develop heterogeneous catalysts based on supported Lewis acid metal requires an initial step where its activity should be evaluated in under homogeneous catalysis conditions [23].

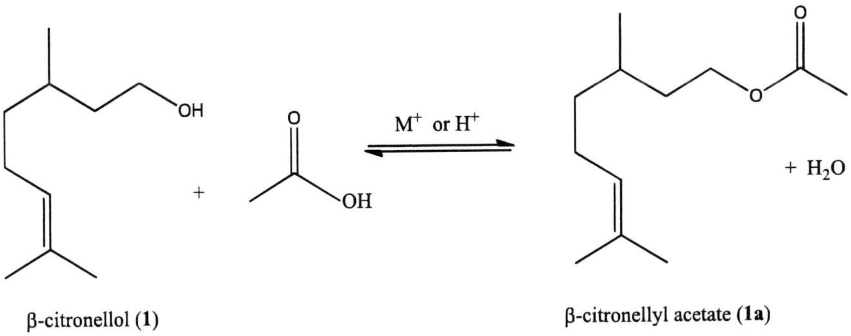

β-citronellol (1)   β-citronellyl acetate (1a)

Figure 1. Lewis or Brønsted acids-catalyzed β-citronellol esterification with HOAc.

Iron(III) salt catalysts have been intensively used in organic synthesis reactions such as oxidative coupling, electrophilic or nucleophilic addition and/ or substitutions, oxidative esterification [24, 25].

Herein, we wish to describe a simple and efficient metal-catalyzed β-citronellol esterification process with HOAc in the presence or absence of solvent.

2. TIN(II)-CATALYZED β-CITRONELLOL ESTERIFICATION WITH HOAC

Tin(II) halides have been catalysts succesfully used on the esterification reactions of glycerol and fatty acids, recently described by our group [26, 27]. We have found that among tin compounds assessed, $SnCl_2$ was the most effective catalyst. Thus, inspired by these findings, we investigated the catalytic activity of $SnCl_2·2H_2O$ on the β-citronellol esterification [28].

Initially, the $SnCl_2$-catalyzed β-citronellol reactions with HOAc were accomplished at different temperatures (ca. 298 to 333 K) in the presence or absence of the catalyst (Table 1).

In the absence of $SnCl_2$, no β-citronellol derivative product was detected regardless temperature used (ca. 298 or 333 K). Conversely, the addition of $SnCl_2$ to the solution triggered two competitive reactions: esterification and oligomerization of β-citronellol.

Table 1. Temperature effects in the $SnCl_2$-catalyzed β-citronellol esterification with HOAc[a] (adapted from ref. 28)

Exp.	T (K)	$SnCl_2.2H_2O$ (mol %)	Conversion (%)	1a Yield (%)	Selectivity (%) 1a	Olig.
1	298	0	0	0	-	-
2	333	0	0	0	-	-
3	298	10	42	37	88	12
4	333	10	21	9	43	57

[a]Reaction conditions: β-citronellol (7.8 mmol), HOAc (15.6 mmol), CH_3CN solution (15 mL), 6 h.
[b](1a) = β-citronellyl acetate; olig. = oligomers.

The temperature had different effects on the conversion and selectivity of the reactions reported in Table 1. While at room temperature a poor conversion and a high selectivity were achieved, conversely, at 333 K temperature, the conversion was two times higher. However, the selectivity was drastically compromised due to the oligomers formation.

The mechanism of these two competitive transformations may be reasonably explained by the Schemes 1 and 2.

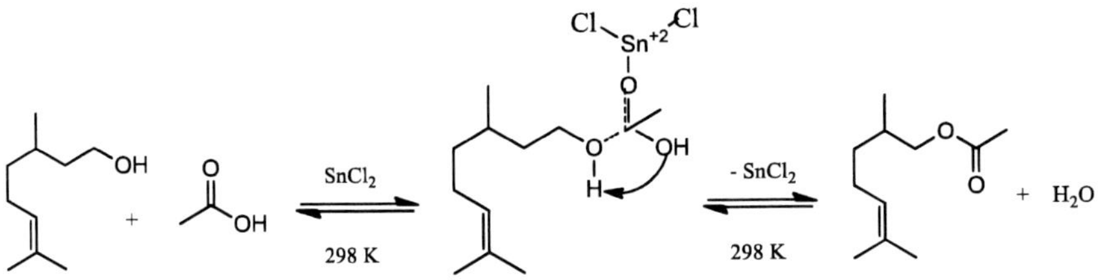

Scheme 1. Mechanism proposal of tin-catalyzed esterification of alcohols with HOAc (adapted from ref. 28).

The presence of the SnCl₂ catalyst in the solution activates the carbonyl group of acetic acid, favoring thus its attack by hydroxyl group of β-citronellol. After the formation of tin intermediate showed in the Scheme 1, an elimination step of water molecule provides the β-citronellyl acetate.

Measurements of pH were carried out of the pure β-citronellol and after the addition of SnCl₂·2 H₂O. It was demonstrated that H_3O^+ cations are formed even in the absence of the HOAc. Indeed, it was found that pH values decreasing from 4.7 to 0.22, after an addition of 7.5 mol % of tin(II) chloride.

Table 2. pH Values before and after addition of tin(II) chloride[a]

System	pH value
β-citronellol	4.70
β-citronellol/ SnCl₂ H₂O	0.22

[a]β-citronellol (3 mL); SnCl₂ (7.5 mol %)

The SnCl₂ undergone alcoholysis reaction in the presence of β-citronellol; this fact was confirmed by the pH measurements of the pure citronellol and with SnCl₂ (Table 2). So, the reaction showed in the Equation 1 may clarify the lowering of the pH values (Scheme 2), due to formation of HCl.

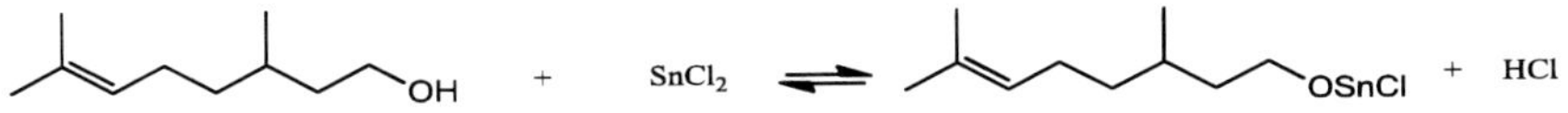

Scheme 2. Alcoholysis reaction of the SnCl₂ with β-citronellol.

Moreover, even at low temperatures, we have found that a stoichiometric amount of dehydration products was formed by the simple addition of SnCl₂·to the β-citronellol. These products were identified through GC-MS analysis as myrcene and 3,7 dimethyl-1,6 octadiene.

Therefore, we assume that the oligomers could be formed throughout the β-citronellol esterification reactions by the action of SnCl$_2$ catalyst, which is able to promote two consecutive reactions (i.e., dehydration, oligomerization; Scheme 3). It is worth mentioning that HCl produced during the hydrolysis can also promoting these two reactions.

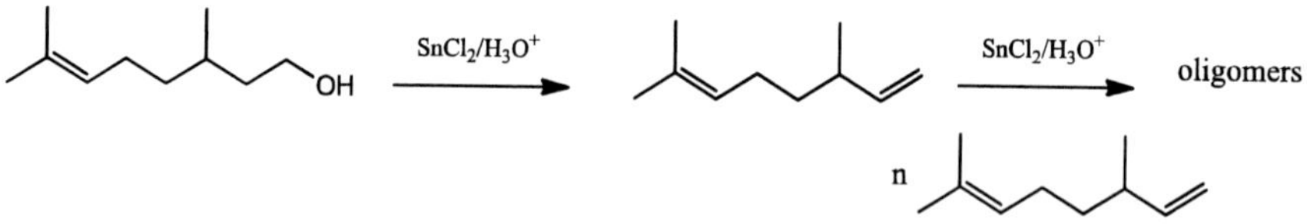

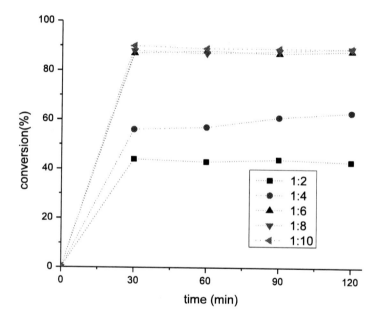

Scheme 3. Mechanism proposal of SnCl$_2$-catalyzed dehydration followed by the oligomerization of β-citronellol.

Figure 2. Effects of reactants stoichiometry on conversion of the SnCl$_2$-catalyzed β-citronellol esterification reactions with HOAc[a] (adapted from ref. 28).
[a]Reaction conditions: β-citronellol (13.1 mmol), HOAc (ranging from 1:2 to 1:10 proportions), SnCl$_2$·2H$_2$O (10 mol %), 298 K, 3 h.

Oligomers have high molar weight and are not detectable through GC analysis. For these reason, they were calculated by the difference between the GC peaks area of detected products (i.e., β-citronellyl acetate) and the GC peak area of consumed substrate (β-citronellol). Before, to calculation it was require to dermining the response factor of product in relation to the substrate, via co-injection of both samples at the same concentrations.

The β-citronellol esterification with HOAc is a reversible reaction. Therefore, a reactant excess may shift the equilibrium toward product formation. However, the Figure 2 shows that at an equal or higher molar ratio than 1:6, no significant increase was attained in the reactions of β-citronellol with HOAc.

On the other hand, selectivity and yields of the reactions were kept almost constant and were not influenced by increase of the HOAc: β-citronellol proportion (Table 3).

The catalyst concentration does not change the reaction equilibrium; nonetheless, we have found that working at the conditions studied, an increasing on the catalyst load resulted in a higher conversion (Figure 3).

Table 3. Temperature effects in the SnCl$_2$-catalyzed β-citronellol esterification with HOAc[a] (adapted from ref. 28)

Exp.	β-citronellol: HOAc molar ratio	Conversion (%)	Yield (1a)[b] (%)	Selectivity (%)[b] 1a	Olig.
1	1:2	43	42	97	3
2	1:4	64	63	98	2
3	1:6	85	84	99	1
4	1:8	86	85	99	1
5	1:10	86	85	99	1
6[c]	1:10	7	7	97	3

[a]Reaction conditions: β-citronellol (13.1 mmol), HOAc (ranging from 1:2 to 1:10 proportions), SnCl$_2$·2H$_2$O (10 mol %), 298 K, 3 h.

[b](1a) = β-citronellyl acetate; olig. = oligomers determined by mass balance via GC analysis; yield determined via GC analysis.

[c]No catalyst.

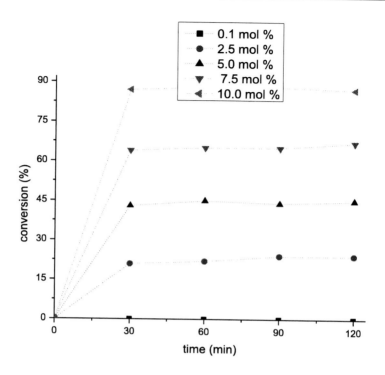

Figure 3. Effects of catalyst concentration on conversion of the β-citronellol esterification reactions with HOAc (adapted from ref. 28)[a]
[a]Reaction conditions: β-citronellol (13.1 mmol), HOAc (78.45 mmol), 298 K, 2 h (adapted from ref. 28).

The reaction selectivity remained almost unaltered when the reactions were carried out with a higher catalyst concentration. Nevertheless, the reactions with low $SnCl_2$ load achieved lower conversion. It means that the reactions do not reached the equillibrium within the 2 hour assessed.

Another important issue is that the profile of the kinetic curves displayed in Figure 3 suggests that after initial period of reaction, the catalyst apparently suffered a deactivation. It was noted regardless the concentration investigated. We suppose that the cause for this deactivation is that Sn^{+2} cations may be being hindered by the remaining β-citronellol molecules; so, after the beginning of the reaction, it result in an alkyl alkoxide of Sn(II), an intermediate that is less active than initial catalyst. It should be better evaluated through future works by our research group.

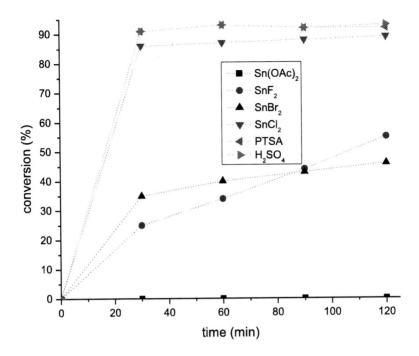

Figure 4. Effects of nature of catalyst on conversion of the β-citronellol esterification reactions with HOAc (adapted from ref. 28)[a]
[a]Reaction conditions: β-citronellol (13.1 mmol), HOAc (78.45 mmol), catalyst (10 mol %), 298 K, 2 h.

The results obtained carrying out the β-citronellol esterification reactions with HOAc in the presence of different Lewis (i.e., Sn(II) salts) and Brønsted acid catalysts are displayed in the Figure 4 and Table 4.

It was found that among the tin salts, SnCl₂ achieved highest conversion (Figure 4). Truly, in addition to the stannous chloride, the more active catalysts were the Brønsted acids (i.e., PTSA and sulfuric acid).

It was expected that if an anionic ligand is present in the tin(II), depending on its donor or withdrawing of electrons characther, it would to do the tin catalyst more or less efficient to activate the carbonyl group of HOAc, thus resulting in a higher or lower conversion. Nonetheless, the solubility of salts as SnF₂ and SnBr₂ was very low and, probably, compromised its catalytic efficiency. As consequence, became difficult linking the activity of tin catalysts to the anion's nature coordinated to the Sn(II).

On the other hand, even that the catalyst has been totally soluble like the case of tin(II) acetate, the reactions achieved the lowest conversion. It can be attributed to the ligand acetate, that is a bidentate ligand and probably hindered the coordination of Sn(II) to the carbonyl group of HOAc and its further activation, so, resulting in a poor conversion (*ca.* < 5%).

Table 4. Effects of catalyst nature on the β-citronellol esterification reactions with HOAc (adapted from ref. 28)[a]

Run	Catalyst	Conversion (%)	β-citronellyl acetate yield (%)
1	H_2SO_4	95	94
2	PTSA	95	94
3	$SnCl_2 \cdot 2H_2O$	86	84
4	SnF_2	65	64
5	$SnBr_2$	47	46
6	$Sn(CH_3COO)_2$	3	0

[a]Reaction conditions: β-citronellol (13.1 mmol), HOAc (78.45 mmol), catalyst (10 mol %), 298 K, 2 h.

The tin(II) acetate was the only catalyst assessed that do not converted the β-citronellol to β-citronellyl acetate. Although different conversions have been obtained, all the reactions with diverse catalysts were highly selective toward of β-citronellyl acetate.

3. Pd(OAc)₂-Catalyzed β-Citronellol Oxidation with Molecular Oxygen

Herein, the main objective was evaluate the catalytic performance of Pd(OAc)₂ in toluene solutions, containing nitrogen and anionic bases, which may act as ligands as well as exogenous base, in the oxidation reactions of β-citronellol by molecular oxygen (Figure 5).

This system was highly selective to oxidizing β-citronellol to β-citronellal in toluene solutions, where the molecular oxygen, an oxidant of minimal environmental impact was used. The effects of different nitrogen ligands and anionic bases on the activity and selectivity of the Pd(OAc)₂-catalyzed reactions were investigated.

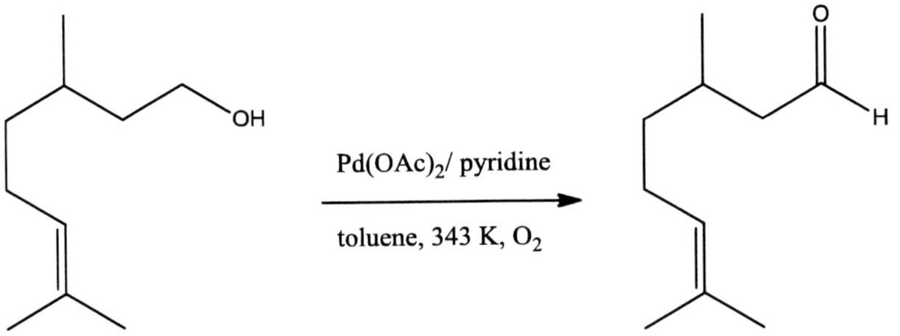

Figure 5. Pd(OAc)$_2$/pyridine-catalyzed β-citronellol oxidation by dioxygen.

In particular, it was found that the system Pd(OAc)$_2$/ pyridine/ K$_2$CO$_3$/ toluene/ O$_2$ was the most effective in promoting the oxidation of β-citronellol, with high conversion, and selectivity to aldehyde higher than 90%.

We have compared our catalytic system to that developed by Uemura's group, which successfully reported an efficient system to oxidizing alcohols (i.e., Pd(OAc)$_2$/pyridine/MS3A/toluene/O$_2$) to aldehydes [29-33]. We have conducted a kinetic study of palladium-catalyzed β-citronellol oxidation in Pd(OAc)$_2$/pyridine/K$_2$CO$_3$ system where the effects of temperature on conversion rate and selectivity were assessed [34].

Initially, we evaluated the impact of nitrogen ligand on activity of palladium acetate catalyst (Figure 6). It is important to note that on this step, the reaction conditions were not optimized to reach high conversions. Our intention was only to select the most efficient nitrogen ligand. Pyridine, sparteine and triethylamine were the nitrogen compounds selected (Figure 6).

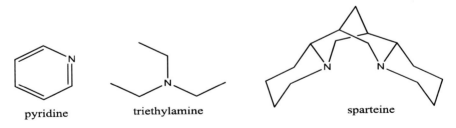

Figure 6. Nitrogen ligands used in palladium-catalyzed oxidation reactions.

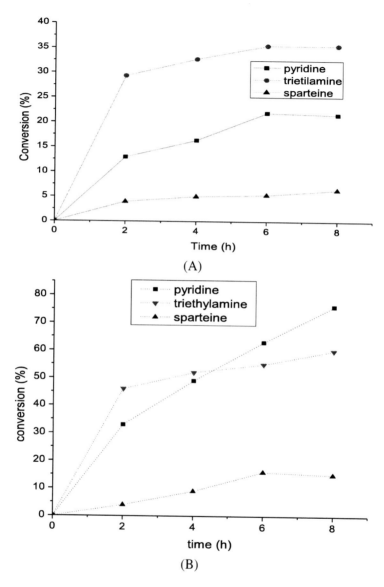

Figure 7. Effect of nitrogen ligands on palladium-catalyzed β-citronellol oxidation by dioxygen in the absence (A) or presence of molecular sieves (B) (adapted from ref 34)[a].
[a]Reaction conditions: β-citronellol (2.75 mmol); nitrogen ligand (5.0 mmol); Pd(OAc)$_2$ (0.05 mmol); reaction volume (10 mL); O$_2$ pressure (0,10 MPa), 343 K, 8h.

All the runs were carried out using excess of nitrogen ligand in relation to the palladium acetate catalyst (1:100 molar ratio), aiming preventing the reduction of Pd(II) to Pd(0) during the reactions. These reaction conditions assure that the nitrogen compounds acts either as ligand and exogenous base.

The nitrogen ligand has as role minimizing the catalyst palladium deactivation, promoting the straight reaction between the palladium reduced species and molecular oxygen [35]. Broad ligands can also modulating the reaction selectivity due to steric hindrance on palladium active site.

Another key aspect is the basicity of nitrogen ligands, which may contribute in the steps of deprotonating of the palladium-hydride intermediates, which are commonly formed during the reactions of palladium-catalyzed alcohol oxidation.

The presence of molecular sieves contribute to removing the water formed in the reaction as well as act as support for that the reactant molecules can react itself. If we keep this points on mind the results showed in the Figures 7 A and B can be reasonably rationalized.

The pkb values of the nitrogen ligands assessed herein are as follow: sparteine (>11.8) [36], pyridine (8.75) and triethylamine (3.35) [37]. On the other hand, sparteine ligand is an alkaloid that has nitrogen atom with high stereochemical hindrance.

In the absence of molecular sieves, the activity of Pd(OAc)$_2$ on the β-citronellol oxidation reactions could be directly associate to the basicity of the ligands; triethylamine was the best ligand for the palladium, as demonstrate in Figure 7A, while sparteine was the worst ligand. Nonetheless, in the reactions carried out in the presence of molecular sieves the pair Pd(OAc)$_2$/pyridine was more efficient than Pd(OAc)$_2$/thiethylamine.

Literature data suggests that MS3A can act a Brønsted base, favoring the deprotonating step, in addition to prevents the formation of palladium bulks, which are responsible by deactivation of palladium catalyst [38]. The combination of these two effects, besides the removal of water from reaction medium explains why the Pd(OAc)$_2$/pyridine or Pd(OAc)$_2$/thiethylamine-catalyzed reactions were more efficient in the presence of MS3A sieves.

The benefic effect was more noticeable in the Pd(OAc)$_2$/pyridine-catalyzed reactions than those with Pd(OAc)$_2$/thiethylamine. The pyridine is π-donating ligand more efficient than triethylamine, therefore, the palladium(0) species were efficiently reoxidized by molecular oxygen. Conversely, the same do not occurred in the reactions with triethylamine, which resulted in the palladium black formation after longer reaction period than 4 hours. Moreover, the ligands triethylamine and mainly sparteine are more bulky than pyridine, so, it has less interaction with the support (i.e., MS3A).

Alternatively to the MS3A molecular sieves, solid carbonates are potential candidates to promoting the palladium-catalyzed oxidation alcohol reactions by molecular oxygen. Its basicity could be useful on deprotonating steps that generally occur in those processes. Consequently, a screening aiming to selecting the carbonate salt more adequate to replacing the MS3A was carried out, and the results are showed in Figure 8.

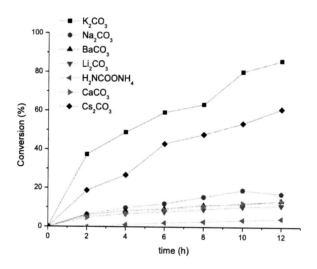

Figure 8. Effect of an anionic base in Pd(OAc)$_2$-catalyzed β-citronellol oxidation by O$_2$ (adapted from ref. 34)[a]
[a]Reaction conditions: β-citronellol (2.75 mmol); Pd(OAc)$_2$ (0.05 mmol); nitrogen ligand (5.0 mmol); anionic bases (2.5 mmol); toluene (10 mL); O$_2$ (0.10 MPa); 333 K.

From the kinetic curves displayed in Figure 8, we can concluded that among carbonates assessed, Cs$_2$CO$_3$ and more mainly K$_2$CO$_3$ enhanced the performance of the Pd(OAc)$_2$/pyridine catalyst. In spite of the low surface area, the efficiency of all the metal carbonates investigated seems be dependent of cation radium. It was found that only for two larger cations a high conversion was achieved, evidence that corroborate this conclusion. So, we have selected the K$_2$CO$_3$ to compare to the MS3A; in presence of these solids, we performed Pd(OAc)$_2$/pyridine-catalysed β-citronellol oxidation reactions (Figure 9).

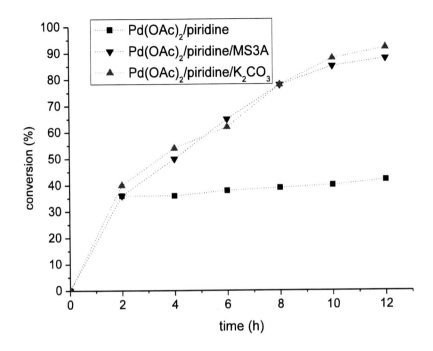

Figure 9. Comparison of systems used on Pd(OAc)$_2$-catayzed oxidation of β-citronellol by molecular oxygen (adapted from ref. 34)[a]
[a]Reaction conditions: β-citronellol (2.75 mmol); Pd(OAc)$_2$ (0.05 mmol); pyridine (5.0 mmol); K$_2$CO$_3$ (2.5 mmol); toluene (10 mL); MS3A (0.5 g); O$_2$ (0.10 MPa); 333 K.

Expectedly, in the absence of MS3A or K$_2$CO$_3$ only a low conversion was achieved. In this case, the palladium deactivation looks have affected the reaction conversion, although no black palladium formation has been observed. Conversely, the reactions carried out with MS3A or K$_2$CO$_3$ achieved high conversions (*ca.* > 90%). It is supporting the conclusion that K$_2$CO$_3$ can effectively replacing the MS3A on these reactions. The low cost and higher commercial availability of K$_2$CO$_3$ are aspects that distinguish it from molecular sieves.

The effect of pyridine load was also investigated and the results showed that an increase of the pyridine: palladium molar ratio did not result in a significant improvement on reaction. However, it is noteworthy that pyridine: palladium loads lower than 4:1 allowed that the catalyst would be deactivated after 4 h of reaction (omitted by simplification in Figure 10).

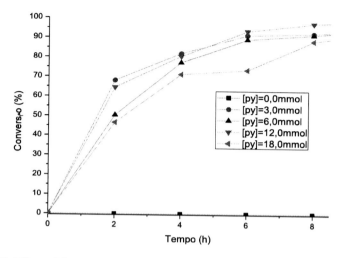

Figure 10. Effect of the pyridine load on Pd(OAc)$_2$-catalyzed β-citronellol oxidation by molecular oxygen (adapted from ref. 34)[a]
[a]Reaction conditions: β-citronellol (2.75 mmol); Pd(OAc)$_2$ (0.050 mmol); K$_2$CO$_3$ (2.5 mmol) toluene (10 mL); 353 K; O$_2$ (0.10 MPa); 8 h[a].

The temperature effects were investigated carrying out reactions at range of 303 to 353 K (Figure 11).

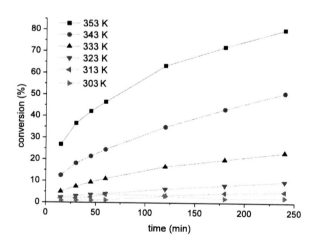

Figure 11. Effect of the temperature on Pd(OAc)$_2$-catalyzed β-citronellol oxidation by molecular oxygen (adapted from re. 34)[a]
[a]Reaction conditions: β-citronellol (2.75 mmol); Pd(OAc)$_2$ (0.050 mmol); K$_2$CO$_3$ (2.5 mmol) toluene (10 mL); O$_2$ (0.10 MPa); 12 h.[a]

The curves displayed in Figure 11 suggest that reaction is endothermic. Mainly for temperatures higher than 333 K, a noticeable increasing on reaction initial rate and conversion were achieved.

Based on literature, we can propose a mechanism for the action of Pd(OAc)$_2$/pyridine catalyst throughout the β-citronellol oxidation by molecular oxygen in toluene [39].

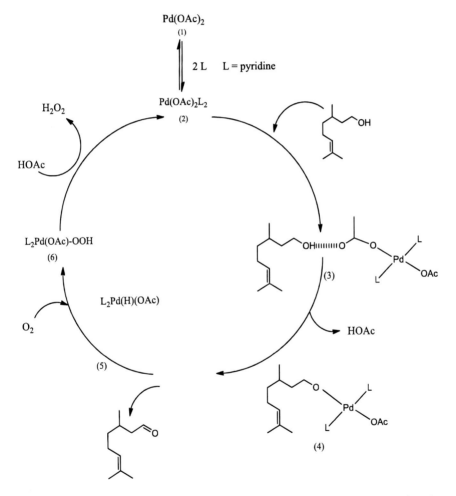

Scheme 4. Possible reaction pathway of the β-citronellol oxidation reaction by dioxygen, catalyzed by Pd(OAc)$_2$/nitrogen ligand (adapted from ref. 34, 39).

Kinetic studies indicate that the Pd(OAc)$_2$ catalyst (Scheme 3) may react with the nitrogen ligand L (L = pyridine) resulting in the L$_2$Pd(OAc)$_2$ intermediate (i.e., intermediate 2, Scheme 4). We think that after the coordination of β-citronellol to this intermediate, which was probably is formed in the medium through hydrogen bonding (intermediate 3, Scheme 4), occur a HOAc elimination step and consequently the formation of Pd(II)-alkoxide (intermediate 4, Scheme 4). Later, the Pd(II)-alkoxide intermediate undergoes β-hydride elimination to form β-citronellal and a palladium-hydride specie (intermediate 5, Scheme 4).

The palladium-hydride intermediate may then undergoes reductive elimination resulting in Pd(0) species (intermediate 6, Scheme 4). The acetate ligand may serve as an internal base in this intramolecular step. The palladium-hydride species can again be converted to an active Pd(II) species by the molecular oxygen, and the Pd(II) complex remains catalytically active throughout the reaction.

4. TIN(II)-CATALYZED SYNTHESIS OF β-CITRONELLYL CARBAMATE THROUGH THE UREA ALCOHOLYSIS

Herein, we have demonstrated that hydrated SnCl$_2$ was an efficient catalyst on the urea alcoholysis with β-citronellol, achieving high conversion and selectivity of β-citronellyl carbamate (i.e., 95 and 90%, respectively) [40]. This novel and selective process provides an inexpensive and attractive alternative to synthesize terpynil carbamate by using of inexpensive and renewable reactants (i.e., urea, β-citronellol) in a single step process.

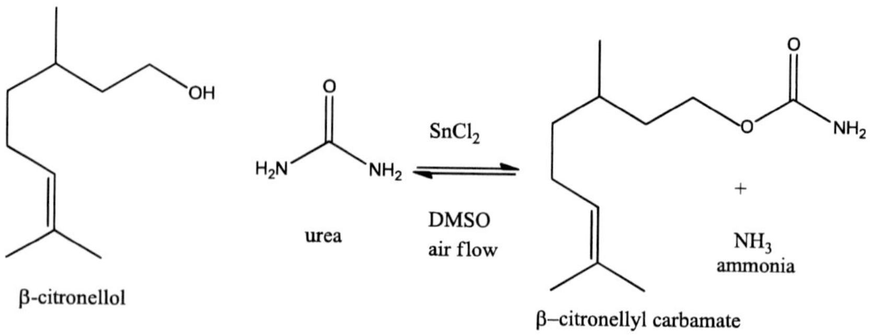

Scheme 5. SnCl$_2$-catalyzed carbamoylation of β-citronellyl with urea (adapted from ref. 40).

We have explored the combination of urea and β-citronellol to synthesize β-citronellyl carbamate, which are pivotally important substrates for production of medicines, pesticides, polymers (i.e., polyurethanes), and protect groups in organic synthesis [41-45].

El-Zemity et al. have synthesized terpinyl carbamates through triethylamine-catalyzed reactions of isocyanate and terpenic alcohols [46]. Although yielding of 70%, one week was required to complete the reaction at room temperature.

Herein, the activity of Sn(II) catalysts was initially assessed in the reactions of urea alcoholysis with β-citronellol. Kinetic curves and main results are summarized in Figure 12 and Table 5, respectively.

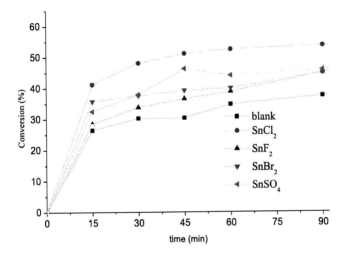

Figure 12. Kinetic curves of tin(II)-catalyzed urea alcoholysis with β-citronellol (adapted from ref. 40)[a]
[a]Reaction conditions: molar ratio of urea to β-citronellol (1:20); temperature (413 K); catalyst (15 mol %) DMSO (15 mL).

Noticeably, the presence of Sn(II) shifted the reaction selectivity toward carbamate formation. However, oligomers were always secondary products. Indeed, the oligomers formation from terpenic alcohols was favored by relatively high reaction temperature. Nonetheless, the addition of Sn(II) cations become the formation of carbamate more effective.

Table 5. Sn(II)-catalyzed urea alcoholysis with β-citronellol (adapted from ref. 40)[a]

Run	Catalyst	Conversion[b] (%)	Selectivity[b] (%)	
			β-citronellyl carbamate	oligomers[c]
1	-	41	20	80
2	SnF$_2$	47	69	31
3	SnCl$_2$	57	85	15
4	SnBr$_2$	49	58	42
5	SnSO$_4$	48	71	29

[a]Reaction conditions: β-citronellol (2.375 mmol); urea (47.5 mmol); temperature (413 K); catalyst (15 mol %) DMSO (15 mL).
[b]Determined by GC.
[c]Calculated from mass balance of reaction.

We suppose that Sn(II) coordination to urea carbonyl group makes its attack by the hydroxyl group of alcohol more favorable. If it is true, it was possible that Sn(II) catalysts containing highly electron withdrawing ligands, may be more active. However, only SnCl$_2$ was totally soluble, hampering a more precise comparison.

Table 6. Effect of SnCl$_2$ concentration in the urea alcoholysis with β-citronellol (adapted from ref. 40)[a]

Run	SnCl$_2$ (mol %)	Conversion (%)	Product selectivity (%)	
			β-citronellyl carbamate	Oligomers
1	0	39	20	80
2	5	54	58	42
3	10	55	57	43
4	15	57	85	15
5	20	30	46	54

[a]Reaction conditions: β-citronellol: urea ratio molar (1:20); temperature (413 K); time (2h); DMSO (15 mL).

The SnCl$_2$ solubility in the reaction solution was dependent of the concentration and limited the assessing on the range of 0 to 20 mol % (Table 6). For concentrations equal or higher than 20 mol %, the catalyst was not completely insoluble. Thus, we verified that a positive effect for conversion and selectivity could be obtained only until 15 mol % of concentration.

The effect of reagents stoichiometry on conversion and reaction selectivity was evaluated in the β-citronelol: to urea molar ratio of 1:5 to 1: 30 (Figure 13).

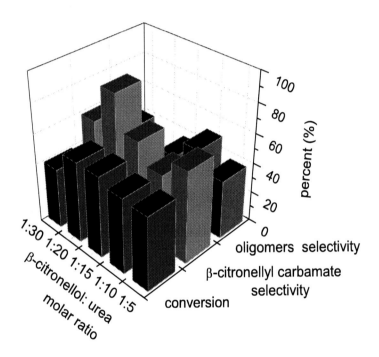

Figure 13. Effect of reactants molar ratio on conversion and selectivity of SnCl$_2$-catalyzed urea alcoholysis with β-citronellol (adapted from ref. 40).

We have found that final conversion was not affected by variation on the reactants proportion (*ca.* maximum of conversion equal to 45 and 55%, in the presence of SnCl$_2$, respectively, 1:20 β-citronellol urea molar ratio). Nonetheless, different of conversion, the reaction selectivity was strongly impacted. The reactions selectivity remarkably improved by the combination of use of a high molar ratio of urea to β-citronellol in presence of SnCl$_2$ catalyst. However, urea solubility decreases when high load is used.

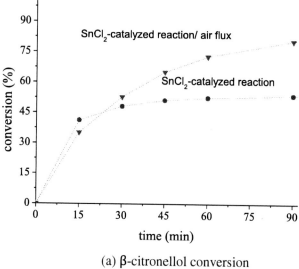

(a) β-citronellol conversion

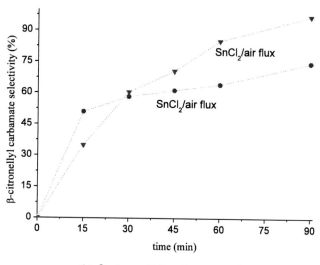

(b) β-citronellyl carbamate selectivity

Figure 14. Effect of air flux on the conversion (a) and selectivity (b) of SnCl$_2$ catalyzed-urea alcoholysis reactions with β-citronellol (adapted from ref. 40)[a].
[a]Reaction conditions: β-citronellol (2.375 mmol); urea (47.5 mmol); DMSO (15 mL); temperature (413 K); SnCl$_2$ (15 mol %).

We find out that the removing of volatile products from the reaction through the air flux shifted the reaction equilibrium toward to higher formation of β-citronellyl carbamate (Figure 14). Remarkably, after 120 minutes of reaction heated to 413 K temperature, and using reactants at 1:20 molar ratio, we have achieved highest β-citronellyl carbamate selectivity (*ca.* 93%) and high conversion (*ca.* 88%), as shown in the Figure 14a. This result is highly superior to that reached in the reactions without flux of air, where although a high β-citronellyl carbamate selectivity (*ca.* 85%), the maximum conversion was ca. 54% (Table 5).

5. OUTLOOK

A simple and straight process to synthesize β-citronellyl carbamate was achieved through $SnCl_2$ catalyzed-urea alcoholysis reactions with β-citronellol. $SnCl_2$ is a commercially available and water tolerant Lewis acid salt and was the most active catalyst to converting β-citronellol to carbamate. In presence of $SnCl_2$, the carbamate selectivity was arguably favoured (*ca.* 85% selectivity, *ca.* 54% conversion). In similar reaction conditions, β-citronellyl carbonate was selectively obtained (*ca.* 80%). The removing of ammonium generated during the carbamoylation reactions using air flux, notably increased the process selectivity. Under that conditions, $SnCl_2$ catalyst efficiently promoted the β-citronellol conversion to β-citronellyl carbamate, achieving high conversion (*ca.* 88%) and selectivity (*ca.* 93%) after a 2 h of reaction at 413 K. The tin-catalyzed carbamoylation process developed herein was more selective and faster than processes synthesis terpenic carbamates reported in the literature.

Finally, we hope that the finding presented in this work may contribute to development of efficient and environmentally benign process for the synthesis of β-citronellol derivatives, which comprises important ingredients for different chemical industries.

ACKNOWLEDGMENTS

The author thanks to the FAPEMIG, Rede Mineira de Quimica, CAPES and CNPq, by financial support.

REFERENCES

[1] C. S. Sell, *The Chemistry of Fragrances: From Perfumer to Consumer.* 2 ed.,V. 2, RSC Publishing, Dorset, 2006, p. 52.
[2] P. Gallezot, *Catal Today. 121*: 76-91 (2007).
[3] C. Chapuis; D. Jacoby, *Appl. Catal.A 221*: 93 (2001).
[4] D. H. Pybus, C. S. Sell, *The chemistry of fragrances.* Paperbacks, RSC, Cambridge, 2001.
[5] A. Corma, S. Iborra, A. Velty, *Chem. Rev. 107*, 2411 (2007).
[6] S. Bhaduri, D. Mukesh, *Homogeneous catalysis: mechanisms and industrial applications.* 1 ed. John Wiley & Sons, New Jersey, 2000. p. 247.
[7] J. L. F. Monteiro, C. O. Veloso, *Topic Catal. 27*: 169 (2004).
[8] W. Schwab, C. Fuchs, F.-C. Huang, *Eur. J. Lipid Sci. Technol. 115*: 3 (2013).
[9] E. J. Lenardao, G. V. Botteselle, F. Azambuja, G. Perin, R. G. Jacob, *Tetrahedron, 63*: 6671 (2007).
[10] M. J. da Silva, A. A. de Oliveira, M. L. da Silva, *Catal. Let. 130*: 424 (2009).
[11] T. Michel, D. Betz, M. Cokoja, S. Volker, F. E. Kühn, *J. Mol. Catal. A 340*: 9 (2011).
[12] M. J. da Silva, A. A. Julio, K. T. dos Santos, *Catal Science Technol. 5*: 1261 (2015).
[13] G. A. Burdock *Fenaroli's Handbook of Flavor Ingredients.*, 5ed. Estados Unidos: CRC Press, 2005. p. 307.
[14] E. J. Lenardao, G. V. Botteselle, F. Azambuja, G. Perin, R. G. Jacob, *Tetrahedron 63*: 6671. (2007).
[15] L. I. R. Ferraz, G. Possebom, E. V. Alvez, R. L. Cansian, N. Paroul, D. de Oliveira, H. Treiche, *Biocatal. Agric. Biotechnol. 4*: 44 (2015).
[16] A. Corma, H. Garcia. *Chem. Ver. 103*: 4307 (2003).
[17] F. Fonteyn, C. Blecker, G. Lognay, M. Marlier, M. Severin, *Biotechnol. Lett. 16*: 693 (1994).
[18] P. Adlercreutz, *Chem. Soc. Rev. 42*: 6406 (2013).
[19] H. F. Castro, E. B. Pereira, W. A. Anderson, *J. Braz. Chem. Soc. 7*: 1 (1996).
[20] R. G. Jacob, *Tetrahedron 63*: 6671 (2007).
[21] P. T. Anastas, L. B. Bartlett, M. M. Kirchhoff and T. C. Williamson, *Catal. Today 55*: 11 (2000).

[22] D. E. Lopez, J. G. Goodwin Jr, D. A. Bruce, E. Lotero, *Appl. Catal. A* *295*: 97 (2005).
[23] J. M. Thomas, R. Raja, *Top. Catal. 53*: 848 (2010).
[24] C. Bolm, J. Legros, J. L. Paih, L. Zani, *Chem. Rev. 104*: 6217 (2004).
[25] C. Rodrigues, R. Fernandez-Lafuente, *J. Mol. Catal. B 66*: 15 (2010).
[26] F. L. Menezes, M. D. O. Guimaraes, M. J. da Silva *Ind. Eng. Chem. Res. 52*: 16709 (2013).
[27] M, J. da Silva, C. E. Goncalves, L. O. Laier *Catal. Lett. 141*:1111 (2011).
[28] M. J. da Silva, A. A. Julio and K. T. dos Santos, *Catal. Sci. Technol. 5:* 1261 (2015).
[29] T. Nishimura, S. Uemura *Catal. Surveys Japan 4*: 31 (2000).
[30] T. Nishimura, T. Onoue, K. Ohe, S. Uemura *Tetrahedron Lett 39*: 6011 (1998).
[31] T. Nishimura, N. Kakiuchi, T. Onoue, K. Ohe, S. Uemura *J. Chem. Soc., Perkin Trans. 1*:1915 (2000).
[32] T. Nishimura, T. Onoue, K. Ohe, S. Uemura *J. Org. Chem. 64*: 6750 (1999).
[33] T. Nishimura, Y. Maeda, K. Kakiuchi, S. Uemura *J. Chem. Soc., Perkin Trans. 1*: 4301 (2000).
[34] D. M. Carari, M. J. da Silva, *Catal. Lett. 142*: 251 (2012).
[35] D. Milstein *Top. Catal. 53*: 915 (2010).
[36] D. Neumann, G. Krauss, M. Hieke, D. Groger *Plant. Med. 48*: 20 (1983).
[37] D. R. Lide (Ed.), CRC Handbook of Chemistry and Physics, 80 ed., CRC Press, Boca Raton, 1999.
[38] B. A. Steinhoff, S. R. Fix, S. S. Stahl, *J. Am. Chem. Soc.124*: 766 (2002).
[39] B. A. Steinhoff, S. S. Stahl, *Org. Lett. 4*: 4179 (2002).
[40] D. M. Chaves, M. J. da Silva, in press, *Catal. Lett.*, 2016.
[41] P. Wang, X. Liu, F. Zhue, B. Yang, A.S. Alshammari, Y. Deng, *RSC Adv. 5*: 19534 (2015).
[42] G. Woods, The ICI Polyurethanes Book, second ed., Wiley, New York, 1990.
[43] P. J. Rice, J. R. Coats, *Pestic. Sci. 41*: 195 (1994).
[44] R. L. Baron, Chapter 17 in R.I. Krieger, W.C. Krieger (Eds.), Handbook of Pesticide Toxicology: Principles and Agents Vol.1, Academic Press, San Diego, 2001, p.1125.

[45] P. G. M. Wuts, T. W. Greene, Greene's Protective Groups in Organic Synthesis, John Wiley & Sons, New Jersey, 2007.
[46] S.R. El-Zemity, *J. Appl. Sci. Res. 2*: 86 (2006).

In: Advances in Chemistry Research. Volume 35 ISBN: 978-1-53610-734-0
Editor: James C. Taylor   © 2017 Nova Science Publishers, Inc.

Chapter 3

DEVELOPMENT AND APPLICATIONS OF A NEW TYPE OF POLYMER-SUPPORTED ORGANOSILICA LAYERED-HYBRID MEMBRANES

Genghao Gong and Toshinori Tsuru[*]
Department of Chemical Engineering, Hiroshima University,
Higashi-Hiroshima, Japan

ABSTRACT

A new type of polymer-supported organosilica layered-hybrid membrane was developed. Using 1,2-bis(triethoxysilyl)ethane (BTESE) as a single precursor, a uniform, thin and perm-selective organically bridged silica active layer was successfully deposited onto a porous polysulfone support via a facile and reliable sol–gel process. These new types of organosilica layered-hybrid membranes were then used for the vapor permeation (VP) dehydration of isopropanol-water (90/10 wt%) solutions, and showed a stable water flux of 2.3 kg/(m² h) and an improved separation factor of about 2500. Moreover, these layered-hybrid membranes also displayed good stability and reproducibility in the reverse osmosis (RO) desalination of a 2000 ppm NaCl solution process,

---

[*] Corresponding Author address: Kagamiyama 1-4-1, Higashi-Hiroshima 739-8527, Japan, Email: tsuru@hiroshima-u.ac.jp.

and showed a stable and high degree of water permeability (approximately 1.2×10^{-12} m^3 m^{-2} s^{-1} Pa^{-1}) with salt rejection that was competitive (96%) with conventional processing. It shows that the separation performances of these polymer-supported organosilica layered-hybrid membranes have been equal to, or even better than, many ceramic-supported membranes.

Keywords: organosilica, layered-hybrid membrane, vapor permeation, reverse osmosis

INTRODUCTION

Membrane-based separations have received considerable amount of attention in chemical industries due to their low energy consumption, low operating costs and steady-state operation [1]. Currently, the improvements in membrane separation technology are mainly focused on the development of advanced separation membranes and the optimization of different membrane processes through process integration [2, 3]. Generally, the separation performance of a membrane is intrinsically determined by the structure and characteristics of the membrane materials [4]. Therefore, the fabrication of advanced separation membranes is one of the most important factors and is attracting great interest.

In recent years, a potential class of organosilica materials with bridging organic groups was developed and applied to various membrane separation processes due to their excellent molecular sieving abilities, adjustable pore sizes, and their superior levels of thermal and chemical resistance [5-7]. For example, using a bridged precursor, 1,2-bis (triethoxysilyl)ethane (BTESE), a promising organosilica membrane was fabricated on a porous ceramic support via a sol–gel process. The BTESE-derived silica membranes had shown an unparalleled stability under a continuous process for dehydration of n-butanol at 150°C over 1000 days [8]. Also, Kanezashi et al. fabricated a BTESE membrane via a "spacer" technique to allow tuning of the silica networks, which also improved the hydrothermal stability and high hydrogen permeability in gas separations [9]. Meanwhile, Xu et al. applied BTESE membranes for the desalination of a 2000 ppm NaCl aqueous solution via a RO process, and the membrane showed excellent chlorine stability for a wide range of chlorine concentrations (~35,000 ppm·h), as well as a superior molecular sieving ability [10]. Moreover, these BTESE membranes are also stable in acid solutions (pH = 2) including both inorganic (nitric) and organic (acetic) acids [11, 12]. Over the past decade, however, almost all of these organosilica membranes were fabricated on flat or tubular ceramic supports [13-15] as shown in Figure 1 (a), which limits the applications of them due to the high cost of the support materials, complex fabrication processes and poor reproducibility, which results in scale-up difficulties [16].

Compared with the ceramic supports, organic polymeric supports are cheap, easily available in large amounts and widely used in industrialization [17]. Recently, therefore, a new concept in the development of advanced separation membranes is now being proposed—the fabrication of layered-hybrid membranes consisting of a high-performance inorganic separation layer deposited onto a porous flexible polymeric support, as shown in Figure 1 (b). For example, Nair et al. attempted to fabricate modified mesoporous silica and zeolitic imidazolate framework (ZIF) membranes on polymeric hollow fibers [18, 19], which they applied to gas separation and to the separation of hydrocarbons by pervaporation, respectively. Reportedly, the separation performances of these membranes have been equal to, or even better than, those of previous ceramic-supported membranes. Subsequently, Fernando et al. reported the fabrication of a ZIF membrane on a flat porous polysulfone (PSF) support via in-situ synthesis. This membrane also showed high H_2/CH_4 (10.5) and H_2/N_2 (12.4) selectivities in a separation test of gas mixtures [20]. Moreover, Ngamou et al. proposed the use of an expanding thermal plasma chemical vapor deposition (ETP-CVD) technique to fabricate a more promising BTESE-derived silica layer on a porous polyamide-imide substrate for n-butanol dehydration [21]. This novel approach avoids the calcination step and the obtained layered-hybrid organosilica membrane showed separation performance comparable to that of ceramic membranes. An alternative approach to fabricating organosilica layers on porous polymeric supports was proposed by Gong et al. [22]. Using a facile, sol–gel spin coating process, a thin and uniform BTESE-derived silica layer was successfully deposited onto a porous polysulfone support. The obtained layered-hybrid organosilica membranes were applied to the vapor permeation (VP) dehydration of an isopropanol–water solution [23] and to the reverse osmosis (RO) desalination of a NaCl solution, and showed good stability and reproducibility [24, 25].

In this chapter, by summarizing our previous works [22-25], a new concept in the development of advanced separation membranes is introduced—the fabrication of layered-hybrid membranes consisting of a high-performance organosilica active layer deposited onto a porous polymeric support. The fabrication process, characteristics and applications of this new type of polymer-supported organosilica layered-hybrid membrane will be discussed in detail.

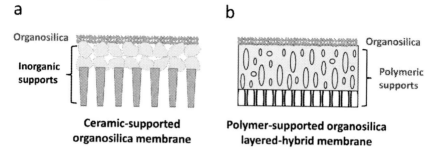

Figure 1. Schematic diagram of the conventional ceramic-supported organosilica membrane (a) and polymer-supported organosilica layered-hybrid membrane (b)

METHOD AND DISCUSSION

1. Preparation of a Polymer-Supported Organosilica Layered-Hybrid Membrane

A polymeric sol, 1,2-bis (triethoxysilyl)ethane (BTESE: $(EtO)_3SiH_2C-CH_2Si(OEt)_3$), was synthesized via hydrolysis and polymerization in a mixture of water, HCl and 1-propanol. First, a given mass of BTESE was mixed with 1-propanol, and then distilled water and HCl were added dropwise to the mixture under continuous stirring. The molar ratio of the BTESE/H_2O/HCl mixture was 1:60:0.1 with 5.0 wt% of BTESE. This mixed solution was stirred for 1.5 h in a closed glass bottle at 60°C.

Figure 2. Schematic illustration of the preparation process for BTESE/NTR layered-hybrid membranes [22].

The fabrication of the NTR-supported organosilica layered-hybrid membrane was carried out via a sol–gel, spin-coating process as shown in Figure 2. Briefly, a sulfonated polyethersulfone (SPES) nanofiltration membrane (NTR-7450, Nitto Denko) with a 2.5 cm diameter was placed on a macroporous stainless substrate (pore diameter: 100 μm, porosity: 50%). Approximately 200 μL of BTESE sol was subsequently dispensed onto the NTR support. The rotation speed was accelerated from 0 to 5000 rpm in 5 s and then held there for 30 s. The spin-coated samples were dried for 10 min at room temperature. This spin coating process was repeated twice. Finally, the obtained samples were heated at 120°C for 20 min.

2. Characterization of a BTESE/NTR Layered-Hybrid Membrane

A typical SEM image of the cross-section of the BTESE/NTR layered-hybrid membrane is shown in Figure 3. A dense, uniform and approximately 200 nm thick BTESE-derived silica active layer was deposited onto NTR support surface (see insert Figure 3). In addition, small amounts of BTESE sols penetrated into NTR support surface during the spin-coating process, leading to the formation of an interlocked structure at the interface between the BTESE layer and the NTR support. This interlocked structure formed at the interface between them helped avoid the formation of pinholes or cracks in the BTESE top layer, and also enhanced the interfacial adhesion between them via the mechanical interlocking effect [23, 26].

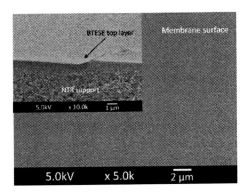

Figure 3. SEM image of the BTESE/NTR layered-hybrid membrane. The insert is a cross-section of the membrane that shows the NTR support and a BTESE-derived silica top layer [25].

Figure 4 (a) shows the ATR-FTIR spectra of the BTESE, the NTR support, and the BTESE/NTR layered-hybrid membrane. The Si–O–Si and Si–OH groups in the BTESE were easily recognized by the broad and strong peaks at 1000–1200 cm^{-1} and 900–950 cm^{-1}, respectively. Compared to the spectrum of NTR support, a new peak that appeared at 900–950 cm^{-1} in the spectrum of the BTESE/NTR membrane was assigned to the Si–OH group. In addition, in the spectrum of this layered-hybrid membrane, a broad but intensive absorption band is apparent at 1000–1200 cm^{-1}, and was assigned to Si–O–Si, though it overlapped the peaks of the NTR support. These results confirmed that a BTESE layer had formed on the NTR support. Moreover, the thickness of the BTESE separation layer was further confirmed by the XPS depth profile analysis, and the results are shown in Figure 4 (b). The concentration of Si atoms that was attributed to the BTESE layer was detected on membrane surface and showed a sharp decrease in depth at approximately 150 nm. This indicated that the effective BTESE active layer had a thickness of about 150 nm. Meanwhile, the concentrations of Si and S atoms showed a sharp decrease and a slight increase, respectively, at a depth from 30 to 150 nm. This atomic mixing of Si and S suggests that an interlocked structure was formed between the BTESE layer and the NTR support due to the penetration of small amounts of BTESE sols.

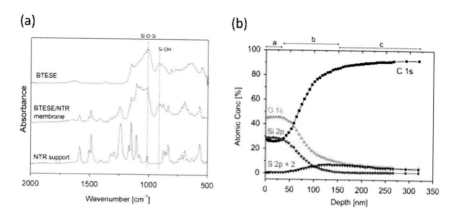

Figure 4. The ATR-FTIR spectra (a) of the BTESE, the BTESE/NTR membrane, and the NTR support, and XPS depth profile (b) of C 1s, Si 2p, O 1s, and S 2p in the BTESE/NTR layered-hybrid membrane [23].

3. Vapor Permeation Dehydration of Isopropanol-Water Solutions

The vapor permeation (VP) performances of BTESE/NTR membranes prepared with different spin-coating cycles are displayed in Figure 5 (a). Clearly, with an increase in number of spin-coating cycles, the separation factor increased while the water flux was slightly decreased. By comparing the membranes spin-coated for 1 and 2 cycles, the separation factor was greatly increased while the water flux was slightly decreased after 2 spin-coating cycles. We speculated that a discontinuous BTESE separation layer might have formed on the support surface after only 1 spin-coating cycle. Elbaccouch et al. also reported that multiple spin-coating cycles were conducive to the formation of a dense, pinhole-free and complete thin film while a single-coating cycle resulted in an incomplete film [27]. However, it is worth noting that the separation factor after 3 spin-coating cycles was only slightly increased (from 2400 to 2600), while the water flux decreased from 2.3 to 1.8 kg/(m^2 h) compared with 2 spin-coating cycles. This indicates that a continuous BTESE layer with high separation performance was obtained with at least 2 spin-coating cycles, whereas a third spin-coating process merely resulted in an increased thickness of the BTESE layer, which is generally associated with a decrease in water flux. In addition, it should be noted that a membrane spin-coated for 5 cycles often showed relatively high water flux and low separation factors. One possible reason could have been that 5 spin-coating cycles led to a thicker BTESE layer on the support, which was prone to crack formation. Another reason could have been that the thick BTESE layer on the flexible polymer support was fragile and more liable to rupture because of a slight deformation in the support during the membrane assembly process or in vacuum. Therefore, the results showed that a uniform, continuous BTESE layer with high separation performance was formed on NTR supports after at least 2 spin-coating cycles.

In addition, Figure 5 (b) shows the time course of VP performance for BTESE/NTR membranes prepared with 2 spin-coating cycles. The VP performance of this membrane gradually became stable after a moderate change over several hours. It is worth noting that this membrane showed a relatively stable water flux and a slow decrease in the IPA flux during the initial 8 h, which caused a gradual increase in the separation factor. This was probably due to the presence of Si–OH groups in the BTESE structure, which were caused by a low degree of condensation reaction in the silanol groups at relatively low curing temperatures. Herein, the adsorption or reaction of IPA molecules with the Si–OH groups might have occurred on the pore surface of the BTESE layer, leading to a reduction in pore sizes, which could prevent IPA molecules from permeating the membrane [28].

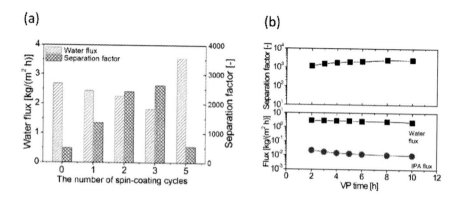

Figure 5. (a) The water flux and separation factor of BTESE/NTR membranes coated with different spin-coating cycles and cured at 150°C for a water/IPA vapor (IPA: 90 wt%) at 105°C. (b) Time course of VP performances for BTESE/NTR membranes prepared with 2 spin-coating cycles [23].

Table 1. Summary of PV and VP dehydration performance of IPA/H₂O vapors for ceramic-supported and polymer-supported BTESE-derived silica membranes [23]

Membrane type	Calcination temperature [°C]	Water flux [kg/(m²h)]	Separation factor [-]	Membrane process (Temp, °C)	Feed concentration [wt%]	Ref.
BTESE/ Al₂O₃	100	2.51	130	PV (75)	90%	[28]
	200	2.86	810			
	300	3.4	4370			
BTESE/PSF	150	1.6	315	VP (105)	90%	[22]
BTESE/ NTR-7450	150	2.3	2500			This chapter

Table 1 summarizes the PV and VP separation performance of IPA aqueous solutions of ceramic/polymer-supported BTESE membranes. Compared with ceramic-supported BTESE membranes, the separation performance of the BTESE/NTR membrane was similar, or even better, at low calcination temperatures (100–200°C). In addition, both water flux (2.3 kg m⁻² h⁻¹) and separation factors (2500) for the BTESE/NTR membrane were much better than those of the BTESE/PSF membrane (a layered-hybrid membrane was prepared on polysulfone (PSF, pore size: 30–50 nm) support).

Compared with the PSF supports with large pore sizes, NTR supports have an ultrathin SPES skin layer and small pore sizes (1.4 nm), which are similar to those of the intermediate layer of ceramic membranes, enabling the formation of a defect-free top layer. This suggests that the choice of a more appropriate polymer support would greatly improve the performance of polymer-supported organosilica layered-hybrid membranes.

4. Reverse Osmosis Desalination of the BTESE/NTR-7450 Layered-Hybrid Membrane

Figure 6 shows the time course for the results of reverse osmosis (RO) performed on a 2000-ppm NaCl solution through BTESE/NTR-60, 120 and 150 membranes which were heat-treated at 60, 120 and 150°C, respectively. It is clear that both BTESE/NTR-120 and BTESE/NTR-150 membranes showed a stable water permeability and a slight increase in salt rejection over time. However, these results show that the salt rejection by the BTESE/NTR-60 membrane decreased rapidly while water permeability increased gradually over the first several hours. This indicates that the RO performance of the BTESE/NTR heat-treated at 60°C was not stable over time. This is probably because the low heat-treatment temperature caused a low degree of condensation reaction by the silanol group, which caused the formation of a loose silica structure [28] that resulted in a lowering of the mechanical strength of the silica networks. Hence, this low-quality BTESE layer was prone to detachment or to peeling off from its NTR supports during the RO process as a result of the shear stress caused by the fluid flowing over the membrane surface via stirring.

The membrane characteristics such as permeate flux and molecular weight cut-off (MWCO) reflect molecular separation performance and were evaluated using several neutral solutes of different molecular weights (MW): ethanol (46), isopropanol (60), glucose (180), maltose (342), raffinose (504), and cyclodextrin (973) at 1.5 MPa and 25°C. As shown in Figure 7, all BTESE/NTR membranes including the BTESE/NTR-60, BTESE/NTR-120 and BTESE/NTR-150 membranes showed a high degree of rejection for neutral solutes with a low molecular weight, because of the presence of a BTESE separation layer compared with a NTR support. However, compared with a BTESE/NTR-60, both the BTESE/NTR-120 and BTESE/NTR-150 membranes showed a higher rate of rejection for neutral solutes with a low molecular weight such as isopropanol (isopropanol: 66% (BTESE/NTR-120) and 70% (BTESE/NTR-150); glucose:>95%; maltose:>99%). These results suggest that the pore size of the layered-hybrid membranes might be different according to different heat-treatment temperatures. It should be noted that the smaller pore size of the BTESE layer on the NTR support was obtained by heat-treatment at a higher temperature, which caused a higher rejection and a lower rate of water permeate flux. The MWCOs defined at 90% rejection for BTESE/NTR-120 and BTESE/NTR-150 membranes were approximately 120 and 130, respectively. In addition, the volume permeate fluxes of different solutes for the BTESE/NTR layered-hybrid were constant, indicating that the permeate flux was not dependent on the types of solutes due to low feed concentration.

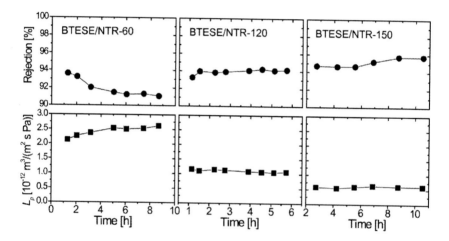

Figure 6. Time courses of the RO performance of BTESE/NTR membranes during the desalination of a 2000-ppm NaCl solution (The BTESE/NTR-60/120/150 membranes were prepared via heat-treatments at 60, 120 and 150 °C, respectively) [24].

Finally, Figure 8 shows the water permeability and salt rejection of a BTESE/NTR-120 membrane as a function of operating time for the RO desalination of a 2000-ppm NaCl solution during a long-term stability test. The water permeability (round 1.2×10^{-12} m^3 m^{-2} s^{-1} Pa^{-1}) and NaCl rejection (about 95%) were almost constant during this continuous RO process—even after more than 160 h. This suggests that the BTESE/NTR layered-hybrid membranes will be applicable in a continuous RO process with a good long-term stability.

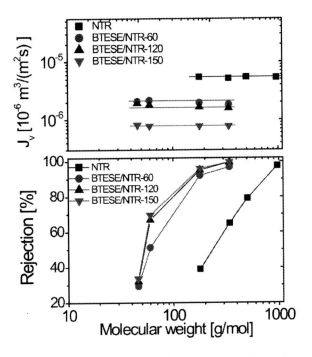

Figure 7. Volume flux and rejection as a function of the molecular weights of neutral solutes for NTR and BTESE/NTR membranes (the BTESE/NTR-60, BTESE/NTR-120 and BTESE/NTR-150 membranes were heat-treated at 60, 120 and 150°C, respectively) [24].

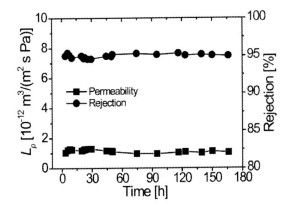

Figure 8. Water permeability and salt rejection of a BTESE/NTR-120 membrane as a function of operating time during the RO desalination of a 2000-ppm NaCl solution [24].

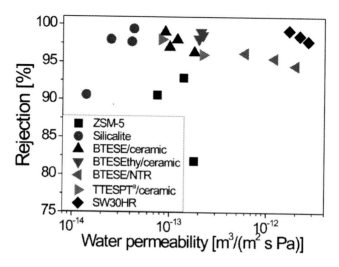

Figure 9. Trade-offs of the desalination performances for zeolite [29] and [30] (25°C, 2.76 MPa, and 0.1 M NaCl), organosilica/ceramic [10, 31] and [32] (25°C, 1.15 MPa, and 2000 ppm NaCl), and commercial aromatic polyamide membranes (SW30HR [33]) (21°C, 2.76 MPa, 2000 ppm-NaCl) RO membranes. (a: an organoalkoxysilane-2,4,6-tris[3-(triethoxysilyl)-1-propoxy]-1,3,5-triazine) [24].

Figure 9 shows the trade-off relationship between NaCl rejection and water permeability for zeolite, organosilica and BTESE/NTR RO membranes. Generally speaking, ceramic-supported organosilica (BTESE, BTESEthy and TTESPT) membranes showed better desalination performance than typical zeolite RO membranes such as ZSM-5 and silicalite. Compared with these organosilica membranes, the water permeability of polymer-supported BTESE membranes was significantly improved and reached 1.2×10^{-12} m^3/(m^2 s Pa), which approximates that of commercial seawater RO membranes (SW30HR), despite a smaller degree of salt rejection due to the low heat-treatment temperature.

CONCLUSION

In this chapter, a novel layered-hybrid membrane consisting of a thin, uniform and defect-free organosilica separation layer on a porous flexible polymeric nanofiltration membrane (NTR-7450) was successfully developed via a facile, sol–gel and spin-coating process. The best vapor permeation (VP) performance was obtained using a perm-selective BTESE layer deposited onto a NTR support with a corresponding water flux of 2.3 kg/m² h and a separation factor of approximately 2500. Moreover, these BTESE/NTR layered-hybrid membranes were first applied to the reverse osmosis (RO) desalination of a 2000 ppm NaCl solution, and displayed good stability in a RO desalination process and excellent reproducibility in the membrane fabrication process. Finally, a high degree of water permeability (approximately 1.2×10^{-12} m³ m⁻² s⁻¹ Pa⁻¹), and a competitive level (approximately 96%) of salt rejection could be achieved.

REFERENCES

[1] Nunes, S. P. and Peinemann, K. V., (2006). Membrane Technology: in the Chemical Industry (Eds.), John Wiley & Sons, Germany.

[2] Li, N. N., Fane, A.G., Ho, W.W., Matsuura, T. (2008). Advanced Membrane Technology and Applications (Eds.), John Wiley & Sons Inc., New Jersey.

[3] Van der Bruggen, B., (2013). Integrated membrane separation processes for recycling of valuable wastewater streams: nanofiltration, Membrane distillation, and membrane crystallizers revisited. *Ind. Eng. Chem. Res.* 52, 10335–10341.

[4] Mohanty, K. and Purkait, M. K., (2011). Membrane Technologies and Applications. New York: CRC Press.

[5] Castricum, H. L., Paradis, G.G., Mittelmeijer-Hazeleger, M.C., Kreiter, R., Vente, J. F., Ten Elshof, J. E., (2011). Tailoring the Separation Behavior of Hybrid Organosilica Membranes by Adjusting the Structure of the Organic Bridging Group. *Adv. Funct. Mater.* 21, 2319–2329.

[6] Castricum, H. L., Sah, A., Kreiter, R., Blank, D. H., Vente, J.F., Johan, E., (2008). Hybrid Ceramic Nanosieves: Stabilizing Nanopores with Organic Links. *Chem. Commun.* 9, 1103–1105.

[7] Xu, R., Ibrahim, S.M., Kanezashi, M., Yoshioka, T., Ito, K., Ohshita, J., Tsuru, T., (2014). New Insights into the Microstructure-Separation Properties of Organosilica Membranes with Ethane, Ethylene, and Acetylene Bridges. *ACS Appl. Mater. Interfaces.* 6, 9357–9364.

[8] van Veen, H. M., Rietkerk, M. D., Shanahan, D. P., van Tuel, M. M., Kreiter, R., Castricum, H. L., ten Elshof, J. E., Vente, J. F., (2011). Pushing Membrane Stability Boundaries with HybSi@ Pervaporation Membranes. *J. Membr. Sci.* 380, 124–131.

[9] Kanezashi, M., Yada, K., Yoshioka, T., Tsuru, T., (2009). Design of Silica Networks for Development of Highly Permeable Hydrogen Separation Membranes with Hydrothermal Stability. *J. Am. Chem. Soc.* 131, 414–415.

[10] Xu, R., Wang, J., Kanezashi, M., Yoshioka, T., Tsuru, T., (2011). Development of Robust Organosilica Membranes for Reverse Osmosis. *Langmuir.* 27, 13996–13999.

[11] Tsuru, T., Shibata, T., Wang, J., Lee, H.R., Kanezashi, M., Yoshioka, T., (2012). Pervaporation of Acetic Acid Aqueous Solutions by Organosilica Membranes. *J. Membr. Sci.* 421, 25–31.

[12] Castricum, H. L., Kreiter, R., van Veen, H. M., Blank, D. H., Vente, J. F., Johan, E., (2008). High-Performance Hybrid Pervaporation Membranes with Superior Hydrothermal and Acid Stability. *J. Membr. Sci.* 324, 111–118.

[13] Agirre, I., Arias, P.L., Castricum, H. L., Creatore, M., Johan, E., Paradis, G.G., Ngamou, P.H.T., Van Veen, H.M., Vente, J.F., (2014). Hybrid Organosilica Membranes and Processes: Status and Outlook. *Sep. Purif. Technol.* 121, 2–12.

[14] Qi, H., Han, J., Xu, N., Bouwmeester, H.J., (2010). Inorganic Microporous Membranes with High Hydrothermal Stability for the Separation of Carbon Dioxide. *ChemSusChem.* 3, 1375–1260.

[15] Qureshi, H.F., Nijmeijer, A., Winnubst, L., (2013). Influence of Sol–Gel Process Parameters on the Micro-Structure and Performance of Hybrid Silica Membranes. *J. Membr. Sci.* 446, 19–25.

[16] Burggraaf, A. and Cot, L., (1996). Fundamentals of Inorganic Membrane Science. *Elsevier Science*: New York, Vol. 4.

[17] Lau, W. J., Ismail, A. F., Misdan, N., Kassim, M. A., (2012). A recent progress in thin film composite membrane: A review. *Desalination.* 287, 190–199.

[18] Jang, K. S., Kim, H. J., Johnson, J. R., Kim, W. G., Koros, W. J., Jones, C. W., Nair, S., (2011). Modified mesoporous silica gas separation membranes on polymeric hollow fibers. *Chem. Mater.* 23, 3025–3028.

[19] Brown, A. J., Johnson, J. R., Lydon, M. E., Koros, W. J., Jones, C. W., Nair, S., (2012). Continuous olycrystalline zeolitic imidazolate framework-90 membranes on polymeric hollow fibers. *Angew. Chem. Int. Ed.* 51, 10615–10618.

[20] Cacho-Bailo, F., Seoane, B., Téllez, C., Coronas, J., (2014). ZIF-8 continuous membrane on porous polysulfone for hydrogen separation. *J. Membr. Sci.* 464, 119–126.

[21] Ngamou, P. H., Overbeek, J. P., Kreiter, R., van Veen, H. M.,Vente, J. F., Wienk, I. M., Cuperusc, P. F., Creatore, M., (2013). Plasma-Deposited Hybrid Silica Membranes with a Controlled Retention of Organic Bridges *J. Mater. Chem. A.* 1, 5567–5576.

[22] Gong, G. H., Wang, J., Nagasawa, H., Kanezashi, M., Yoshioka, T., Tsuru, T., (2014). Fabrication of a layered hybrid membrane using an organosilica separation layer on a porous polysulfone support, and the application to vapor permeation. *J. Membr. Sci.* 464, 140–148.

[23] Gong, G. H., Wang, J., Nagasawa, H., Kanezashi, M., Yoshioka, T., Tsuru, T., (2014). Synthesis and Characterization of a Layered-Hybrid Membrane Consisting of an Organosilica Separation Layer on a Polymeric Nanofiltration Membrane. *J. Membr. Sci.* 472, 19–28.

[24] Gong, G. H., Nagasawa, H., Kanezashi, M., Tsuru, T., (2015). Reverse Osmosis Performance of Layered-Hybrid Membranes Consisting of an Organosilica Separation Layer on Polymer Supports *J. Membr. Sci.* 494, 104–112.

[25] Gong, G. H., Nagasawa, H., Kanezashi, M., Tsuru, T., (2016). Tailoring the Separation Behavior of Polymer-Supported Organosilica Layered-Hybrid Membranes via Facile Post-Treatment Using HCl and HN3 Vapors. *ACS Appl. Mater. Interfaces.* 8, 11060–11069.

[26] Wei, W., Xia, S., Liu, G., Gu, X., Jin, W., Xu, N., (2010). Interfacial adhesion between polymer separation layer and ceramic support for composite membrane. *AIChE J.* 56, 1584–1592.

[27] Elbaccouch, M. M., Shukla, S., Mohajeri, N., Seal, S., (2007). Microstructural analysis of doped-strontium cerate thin film membranes fabricated via polymer precursor technique. *Solid State Ion.* 178, 19–28.

[28] Wang, J., Kanezashi, M., Yoshioka, T., Tsuru, T., (2012). Effect of calcination temperature on the PV dehydration performance of alcohol aqueous solutions through BTESE-derived silica membranes. *J. Membr. Sci.* 415, 810–815.

[29] Li, L. X., Liu, N., McPherson, B., Lee, R., (2007). Enhanced water permeation of reverse osmosis through MFI-type zeolite membranes with high aluminum contents. *Ind. Eng. Chem. Res.* 46, 1584–1589.

[30] Liu, N., Li, L. X., McPherson, B., Lee, R., (2008). Removal of organics from produced water by reverse osmosis using MFI-type zeolite membranes. *J. Membr. Sci.* 325, 357–361.

[31] Xu, R., Kanezashi, M., Yoshioka, T., Okuda, T., Ohshita, J., Tsuru, T., (2013). Tailoring the affinity of organosilica membranes by introducing

polarizable ethenylene bridges and aqueous ozone modification. *ACS Appl. Mater. Interfaces.* 5, 6147–6154.

[32] Ibrahim, S. M., Xu, R., Nagasawa, H., Naka, A., Ohshita, J., Yoshioka, T., Kanezashi, M., Tsuru, T., (2014). Insight into the pore tuning of triazine-based nitrogen-rich organoalkoxysilane membranes for use in water desalination. *RSC Adv.* 4, 23759-23769.

[33] Hatakeyama, E. S., Gabriel, C.J., Wiesenauer, B. R., Lohr, J. L., Zhou, M. J., Noble, R. D., Gin, D. L., (2011). Water filtration performance of a lyotropic liquid crystal polymer membrane with uniform, sub-1-nm pores. *J. Membr. Sci.*, 366, 62–72.

In: Advances in Chemistry Research. Volume 35   ISBN: 978-1-53610-734-0
Editor: James C. Taylor   © 2017 Nova Science Publishers, Inc.

Chapter 4

THE LONG-RANGE ORDER IN VEGETABLE AND PERFUME OILS, BUTTER AND ANIMAL FAT

Kristina Zubow[1],, Anatolij Zubow[2] and Viktor Anatolievich Zubow[1]*
[1]R&D, Zubow Consulting, Germany
[2]Dept. Telecommunication Networks Group, TU Berlin, Germany

ABSTRACT

The long-range order (LRO) in vegetable oils (olive, sunflower, maize) and in some perfume oils (rose, lavender, vermouth, wormwood and lemongrass) was investigated using the gravitational mass spectroscopy (GMS). The LRO of them was compared with that one of butter and pig fat. In all samples, a hierarchy in LRO at the level of clusters and super cluster structures (sub micelles and micelles) prevailed, they were allowed by the gravitational noises (GN) of the universe and built to minimize the potential energy of the adhesive-cohesive interaction between molecules. The cluster formation was satisfactorily described by the first Zubow equation. For the mass range up to 200 million Dalton, the main LRO parameters (average molecular mass,

---

* Corresponding author: Email: kristinazubow@gmail.com.

number of cluster kinds, part of collapsed and expanded forms, cluster distribution) were analyzed.

Keywords: vegetable, perfume oils, butter, pork fat, long-range order, molecular clusters

1. INTRODUCTION

Previously, we have shown that molecules as ensembles of atomic nuclei, secondary and higher structures in polymers (domain concentrations) and clusters in liquids are formed under the influence of local GN that they are resulted from gravitational fluctuations in the sun system and in our galaxy [1].

Natural vegetable oils have to be understood as triglycerides consisting of different higher saturated and unsaturated fatty acids [2] (https://en.wikipedia.org/wiki/Triglyceride). In such liquid mixtures, molecular mass concentrations similar to those in nonanoic acid were assumed to be built [1]. However, the presence of unsaturated groups in the liquid, and twice the length of the aliphatic residues should complicate the construction of the cohesively formed globules leading to three centers: ether groups, aliphatic sequences ($-CH_2-$) and unsaturated groups consisting of ($-CH = CH-$). On the other hand under the impact of GN on oils molecular clusters were allowed to be formed at fixed frequencies of stationary waves in GN, only. Therefore, one can expect some cluster sequences and a cluster mass distribution, reflecting the difference in the chemical building of acids. In the present work the attention will be paid to perfume and edible oils, mainly.

The amorphous structure in, for example, perfume oils is based on terpenes and terpenoids (low molecular substances), so their LRO formation will be somewhat different from that one of large molecular substances. The formation of LRO in vegetable oils under the influence of GN is of important practical interest related primarily to the quality of the resulting product, to a new level of process controlling and to understanding of the anthropogenic pollutants' **impact on the growing process of plants** and finally to the development of a marketing strategy in accordance with taste and cosmetic requirements.

The aim of the present work was to investigate the long-range order in natural oils and fats.

2. MATERIAL AND METHOD

Vegetable oils of food and pharma purity, butter and pork fat (salo, https://en.wikipedia.org/wiki/Salo (food), calorific value ~ 4200 kJ per 100 g, total 87% fat including 32% unsaturated fatty acids, 3% proteins and ca. 8% water) served as investigation objects. The chemical composition of corn (maize) oil was close to that one of sunflower oil (calorific value ~ 3400 kJ per 100 ml, ~10% saturated fatty acids, ~ 82% unsaturated fatty acids including ~ 26% monounsaturated and the rest with multiple double bonds (https://en.wikipedia.org/wiki/Vegetable_oil and Figure 3) with minor deviations of water and some fatty acids. For olive oil (clear and cold pressed, https://en.wikipedia.org/wiki/Olive_oil), the total fatty acid content was equal to 91.6 wt. %. Both for olive and sunflower oil the difference to 100% was assigned to water, mainly. The content of proteins and salts was not more than 0.5%. The chemical composition of essential oils is given in https://en.wikipedia.org/wiki/Essential_oil. We used double-distilled oils with a water content of approx. 5.3%. To reach the thermodynamic balance the substances were aged at 288 ± 3 K for 1 year in a dark box that was protected from mechanical influences and external energy fields (light, noises and electromagnetic radiation) [1]). The temperature deviation during the measurement did not exceed ± 0.5 K and the atmospheric pressure remained constant (103,470 ± 3 N/m^2) during the entire experiment (6 h). For the measurement the GMS spectrometer of the firm "**Aist Handels**-und **Consulting**" GmbH (Germany) was applied where the sensor was placed in the center of a glass cell (20 ml) [1]. The wall impact taken as a systematic error was ignored [1]. The measurement time was equal to 30 s. The GMS spectra were given in the coordinates of cluster masses (m) and their relative shares in the cluster ensemble (f). Two kinds of shock waves were used (1N/m^2 > p > 1N/m^2). The method relies on digital processing of multidimensional signals [1] caused by the weak interaction of shock waves with the molecular clusters in liquids where the masses and oscillation frequencies of clusters in liquids were calculated according to the Zubow equations [1] using the Zubow constant 6.4·10^{-15} N/m (salo, up to log m < 3.5, Figure 5) and 8.2·10^{-15} N/m (oils and fat). To divide between the different cluster forms the collapsed clusters were described with -f and the expanded ones with +f. The bigger the absolute value of f the more individual is the cluster, the smaller is its interaction with the surroundings, its oscillations become more free and on the contrary, the lower f the greater the oscillator interaction with the surroundings and the less the cluster is pronounced as a real particle.

3. RESULTS AND DISCUSSION

3.1. Food and Perfume Oils

Figure 1 shows GMS-spectra of olive and sunflower oils.

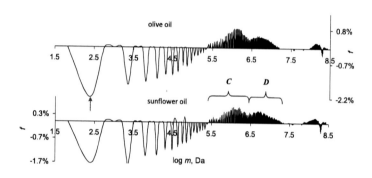

Figure 1. GMS-spectra of vegetable oils, 291 K, $p < 1 N/m^2$, Zubow constant is taken equal to $8.2 \cdot 10^{-15}$ N/m. Arrow indicates a signal being a superposition of small oil clusters and the base water cluster in the collapsed form [1]. Mass clusters (m) are given in Daltons.

The GMS spectra of both olive and sunflower oils were found to be more or less similar under weak shock waves. It was striking, however, that the olive oil contained significantly more water (shown by arrow). Other differences namely two oscillator signals from expanded clusters (log m = 4.57 and 4.82) and a broader *C* + *D* interval were observed for sunflower oil. Here, according to [1], the interval *C* was assigned to sub-micelle cluster ensembles and the interval *D* to micelle ones.

The sub micelle size distribution, for example, in sunflower oil (F_{so}) could be described by the next smoothing function:

$F_{so} = -0.0001 \cdot R^6 + 0.0056 \cdot R^5 - 0.1202 \cdot R^4 + 1.3517 \cdot R^3 - 8.4398 \cdot R^2 + 27.762 \cdot R - 37.574$

The average radius R = 8.36 nm (interval *C*).
For olive oil (F_{oo}), the size distribution followed the equation:

$F_{oo} = 7 \cdot 10^{-6} \cdot R^6 - 0.0003 \cdot R^5 + 0.0052 \cdot R^4 - 0.0469 \cdot R^3 + 0.2172 \cdot R^2 - 0.4576 \cdot R + 0.289$

The average value of R was 8.56 nm (interval *C*).

Further, differences in these oils were visible in the curves of the integral distribution of cluster fractions (W_x), Figure 2.

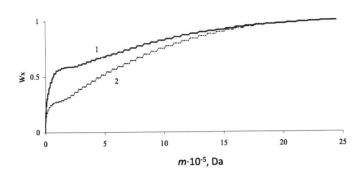

Figure 2. Integral curves of the cluster fraction distribution in sunflower (1) and olive (2) oils.

Figure 3 showed the GMS-spectra of some vegetable oils when they were exposed to strong shock waves from spectrometer. The analysis under strong shock waves is the technological processing conditions very close. Here, can be seen that the spectra of sunflower and maize oils, were almost identical though, both they were something differently from that one of olive oil. This means that under equal technological conditions of vegetable oils LRO is practically the same, namely when the oils were permanently exposed to destroying LRO factors (light, mechanical shocks, sound effects, radiation, etc.). On the other side it was observed that under strong shock waves in the base water cluster of 12 molecules and in the small oil clusters began to dominate the expanded structures (market by arrow), the f value becomes positive [1]. Thus, in sunflower oil appeared small clusters consisting of 6, 15 and 22 molecules with a middle triglyceride mass of 318 Da (signals were marked with 6, 15 and 22, Figure 3). For example, the cluster of 5.5 ~ 6 triglyceride molecules modeled as one sphere may be presented in liquid oil as a uniform oscillator (7353 Hz), for example:

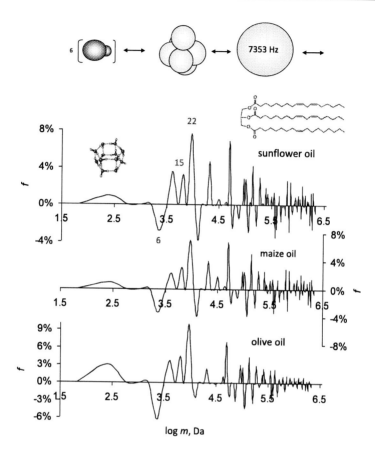

Figure 3. GMS-spectra of some vegetable oils. 295K, $p > 1N/m^2$, Zubow constant is taken equal to $8.2 \cdot 10^{-15}$ N/m. The average sunflower oil molecule structure (ca. 5% palmitic acid, ca. 6% stearic acid, ca. 30 % oleic acid, ca. 59 % linoleic acid) was given above (https://en.wikipedia.org/wiki/Sunflower_oil).

The oscillator with mass of 7002 Da (22 average molecules) gave a signal at 3676 Hz it can be shown as a uniform sphere of 4 oscillators (each at 7353 Hz):

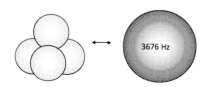

The larger super cluster formations (sub ensembles) were found to be sub micelle and micelle structures themselves built by small clusters.

The comparison of GMS-spectra in the Figures 1 and 3 showed the molecular structuring in oils in the absence of external energy factors (mechanical and electromagnetic influences). When these factors did not act the cluster ensembles were harmonized because of the minimization of the potential energy e.g., at long storage, which is, however, rare in the modern oil technology. Nevertheless, this phenomenon should be taken into account for example in tribology and cosmetics. Thus, liquid vegetable oils are nano colloidal dynamic systems in which exists sub ensembles of similar clusters with related cohesive energies, for example *C* and *D* (Figure 2).

The following table summarized some LRO parameters of oils according to Figure 3.

Table. Main LRO parameters of some vegetable oils under strong shock wave ($p > 1N/m^2$) influence [1], (investigated mass ensembles up to 2 mio. Da). D_c –part of collapsed clusters, $F_{95\%}$ - part of skeletal clusters, M_{GMS} - average cluster mass, in Da, f – part of clusters in expanded or collapsed form with a given mass

Oil	M_{GMS}, Da	f, at 2300 Da	f, at 250 Da	D_c, %	$F_{95\%}$, %
sunflower	423271	-2.8%	1.0%	64	47
maize	426518	-3.4%	0.8%	62	71
olive	290201	-6.3%	3.1%	64	48

Since the main LRO characteristics in vegetable oils were very similar (nearly identical D_c) the average cluster mass strongly differed being for maize and sunflower oils around 420 kDa for olive oil, however, ca. 290 kDa, only. Further, for olive oil, the cluster with 2300 Da strongly densified while the small one (250 Da) highly expanded. On the other side, the part of skeletal clusters in maize oil was much higher than in other oils indicating the high LRO stability. Additionally, significant differences were also analyzed in the area of small clusters. For example, in olive oil the interaction of small clusters (250 ± 20 Da and 2300 ± 300 Da) with their surroundings [1] was less pronounced than in other oils. May be, these differences were due to diverse water contents in the fatty acid clusters changing the cluster shape (expanded or collapsed). These suggestions, however, needed further investigation.

The main components of the perfume oils were represented by terpenes and terpenoids whose structural formula were given in the GMS-spectra (Figure 4). Until the masses of 9200 ± 100 Da, the LRO in the perfume oils was found to be nearly similar. However, in vermouth oil, the large clusters were more pronounced (absolute value of f) than in other oils, where the average mass of all clusters was almost 50% higher. The signal of 154 Da in lavender oil (signal market by an asterisk) was assigned to the collapsed form (-f) of the linalool molecule (154 Da, Cochrane molecular models, scheme I).

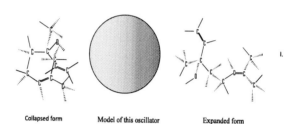

Collapsed form   Model of this oscillator   Expanded form

The signals of the following oscillators in lavender oil were equivalent to the masses of 11, 19 and 31 linalool molecules (https://en.wikipedia.org/wiki/Linalool). Statistically the most likely structures with masses equivalent to 11, 19 and 31 linalool molecules are prototypes of spherical micelles. Except vermouth oil, the integral distribution of cluster fractions in perfume oils, exposed strong shock waves, follows satisfactorily the equation:

$$W_x = (0.125 \pm 0.005) \cdot \ln(m \cdot 10^{-5}) + (0.66 \pm 0.03)$$

The analogous distribution of clusters in vermouth oil was described by an equation of the sixth degree:

$$W_x = -5 \cdot 10^{-6} x^6 + 0.0003 x^5 - 0.0051 x^4 + 0.0465 x^3 - 0.2105 x^2 + 0.491 x + 0.1723,$$ where $x = m \cdot 10^{-5}$.

This curve detected three steps corresponding to three types of sub ensembles (concentrations of similar clusters).

The Long-Range Order in Vegetable and Prefume Oil... 69

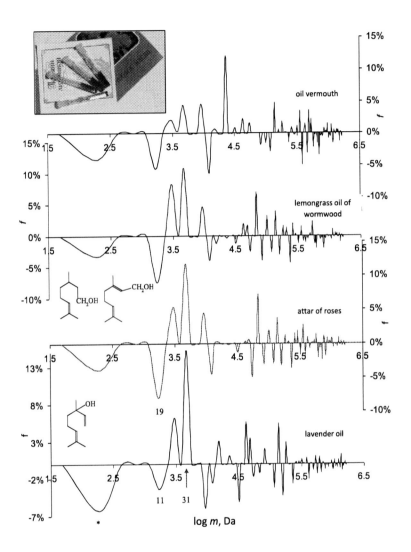

Figure 4. Overview GMS-spectra of some perfume oils (Crimea, Russia). 294 K. Lavender oil, the structure of the basic substance was given in the upper left corner of the spectrum. The structures of the two main components of rose oil citronellol (https://en.wikipedia.org/wiki/Citronellol) and geraniol were shown in the upper left corner of the spectrum for attar of rose. The main LRO parameters for lavender oil: M_{GMS} = 210,822 Da, D_c = 47%, $F_{95\%}$ = 58%, N = 53, for rose oil: M_{GMS} = 224,178 Da, D_c = 49%, $F_{95\%}$ = 64%, N = 53, for wormwood lemongrass: M_{GMS} = 191,530 Da, D_c = 47%, $F_{95\%}$ = 73%, N = 52 and for vermouth: M_{GMS} = 319,142 Da, D_c = 45%, $F_{95\%}$ = 70%, N = 52, p > 1 N/m².

3.2. Animal Fats

For the animal fat, a similar hierarchy of LRO formation at the level of cluster and super cluster structures (sub micelles and micelles, Figure 5) was observed.

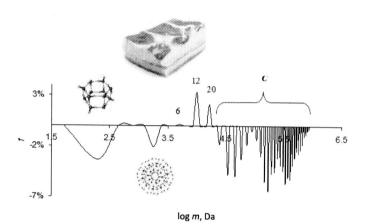

Figure 5. GMS spectrum of fresh pork fat (salo), $p < 1 N/m^2$, Zubow constant 6.4×10^{-15} N/m, 291 K. The modeled base water cluster $(H_2O)_{12}$ and the cluster $(H_2O)_{100}$ were kindly provided by the professors Lenz [6] and Chaplin [3], correspondingly.

Here the distribution of the sub micelle structures (interval C) followed the next smoothing function:

$$F_{af} = 2 \cdot 10^{-5} \cdot R^6 - 0.001 \cdot R^5 + 0.0198 \cdot R^4 - 0.2008 \cdot R^3 + 1.1152 \cdot R^2 - 3.1919 \cdot R + 3.6743$$

The average sub micelle radius R was equal to 8.66 nm.

Let us consider the formation mechanism for the first three sub micelles in the interval C with the masses of 136,163; 160,960 and 187,523 Da. For the first sub micelle, the potential energy minimization was achieved when 3 clusters of 31,167 Da and one cluster of 43,571 Da associated to a new one of 137,071 Da, for the second sub micelle, 3 clusters of 43,571 Da and one of 31,167 Da associated to 161,88 Da. Here the little deviation (< 0.6%) between the modeled and experimental cluster masses were assigned to some molecule fragments that do not participate in the cluster oscillation. The third sub micelle in this series can be satisfactorily modeled as associate of 3 clusters with mass of 43,571 Da and one cluster with mass of 57,877 Da. The rest masses of the sub micelles in the interval *C* also fit well in the harmonization model of LRO as associates of low molecular and sub micelle clusters. The scheme II showed an average statistical structure of the first sub micelle from the interval *C*.

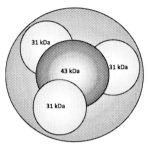

Found 136.1 kDa (Teor. 137 kDa)
II.

In this example, the general principle of the formation of sub micelles can be seen being similar to those in other molecular systems e.g., synthetic polymers and proteins [1]. The formation of huge micellar structures in fats can be understood as a process of sub micelle association. An analogous hierarchy of large clusters' construction from smaller ones was successfully simulated for water by computer programs [3]. This hierarchy is controlled by GN, penetrating our planet [4].

In the literature, there was no information on the LRO construction of animal fats in the subcutaneous layers the general chemical formula was given, only (scheme III):

III.

In this formula, the polar groups are concentrated in the ester groups of the glycerol residue and the nonpolar ones in the aliphatic chains' clusters. It is believed that in a single isolated fat molecule the hydrophobic chains will in the first line cohesively interact with each other to forms allowed by GN. In the presence of other neighboring molecules, the cluster structure may vary according to the nature of the surrounding energy fields. In the study of nonanoic acid - the closest analogue [1], it was found that the molecule formed two cluster kinds of each polar and nonpolar groups. A similar LRO was expected for animal fats.

As visible in Figure 5, there were present the signals of the small water clusters $(H_2O)_{11 \pm 1}$ and $(H_2O)_{100}$ (both in collapsed form [1,3], -*f*), which seem to be overlapped by the weaker signals from aliphatic clusters [1]. Additional information on the state of the water base cluster in oils shall be provided in [1]. Since the water content in fat is not high the signals of all following clusters may be ascribed to fat clusters. Let us study it in detail. Assuming that all three OH groups of glycerol were esterified with stearic acid only (806 Da), than the next 3 signals can be assigned to the clusters of 6, 12 and 20 acids' fragments (marked in Figure 5). For the simplest mass concentration consisting of 6 molecules, the modeled structure was given in scheme IV.

According to this cluster model an oscillation signal from the core, formed from an inside cluster of 6 glycerol residues should be expected however, the signal was not detected to be explained with that it was deleted due to the balance between the expanded and the collapsed form or to that the formation conditions for a single pendulum were not present. The following cluster-12 and cluster-20 were constructed similarly, where in the center the polar ester groups (depending on the polarity of the surroundings) were concentrated and the shell consisted of a layer of aliphatic chains. The signals in the interval *C* were previously identified in proteins, synthetic polymers and vegetable oils, too [1] (Figure 5). Because of the different signal intensities, the clusters were divided into 3 groups - sub ensembles of masses. The ensembles stands for large sub-micelles and even more micelles. Thus, sub micelles and micelle structures similar to those in surfactants (surface active agents [5]) were suggested to exist. Moreover, the cluster character of animal fats ensures the desired skin elasticity (lowering of internal tension), permeability to water, salt solutions and gases that means it is a classical nano colloidal system.

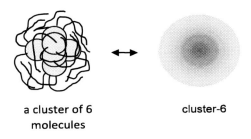

a cluster of 6 molecules   cluster-6

IV.

To investigate this very interesting amorphous matter and the dynamics of its development, which depends on the state of real GN, should become a strategic challenge of natural science in the XXI century.

3.3. Butter

Figure 6 showed the GMS-spectra bio butter (food quality, EU).

As visible the butter GMS-spectra strongly differed from those of vegetable oils and animal fat. While for butter, the small and middle clusters (up to log m <5.7) were in the expanded form and their distribution was harmonically they were for vegetable oils and pork fat both in the expanded and collapsed ones. This LRO harmonization for bio butter was an indication for its long-time storage in special refrigerators.

Let us consider the small cluster BM_8 in more detail. The cluster could be imaged as a micelle in which core the polar ester groups were concentrated and in which shell - the aliphatic chains. This cluster with the mass of 5034 Ds oscillated as a unified structure at 4405 Hz (scheme V).

The sub ensembles built by large sub micelle structures were represented in the collapsed form. The sub ensembles of huge micelle structures on the contrary, in the expanded shape. Similar spectra were obtained for other butter sorts. Analyzing the mass of triglyceride clusters it was shown that the water clusters were not integrated in the triglyceride ones, they were part of the colloid water-oil micro emulsion, only.

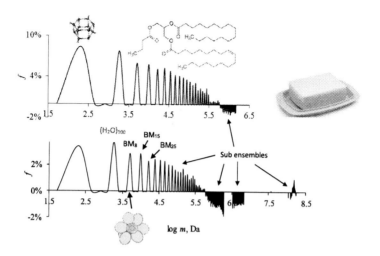

Figure 6. GMS-spectra of bio butter, FRG, https://de.wikipedia.org/wiki/Butter. Water content was up to 16%. Weak shock waves. Average butter molecule (BM) modeled over right (664 Da). The signals of 8 BM were marked as BM_8 (below the model), of 15BM as BM_{15}, of 25BM as BM_{25}. Upper GMS-spectrum for the mass ensemble up to 2 mio. Da, bottom – up to 200 mio. Da.

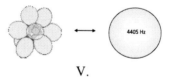

V.

Figure 7 showed the GMS-spectra of another butter sort before and after the removal of water.

As seen in Figure 7, removing the water of butter led first to a changed cluster distribution in the sub ensembles of sub micelle and micelle structures (indicated by arrows) furthermore, to reduced signal intensities of the triglyceride clusters in the expanded form explained that with the water removal the building of expanded triglyceride clusters was inhibited. Therefore, water played a significant role in the formation and stabilization of the huge sub micelle and micelle structures in butter. Here also was shown that the signals from small water clusters were overlapped by those from small triglyceride clusters, where the central cluster of the glycerol residues' trimer gave a signal at 22.222 Hz the entire triglyceride trimer as an oscillator unit however, at 7353 Hz (2315 Da), scheme VI and Figure 6.

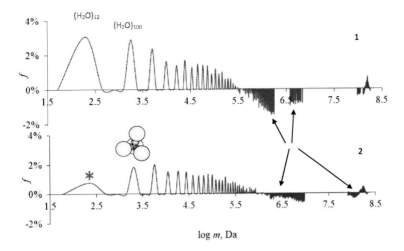

Figure 7. GMS-spectra of butter, FRG, 1 - original commercial product, 2 - butter after water removal. Weak shock waves. The signal from the central cluster as a unit oscillator (22,222 Hz) of three glycerol residues was marked with an asterisk.

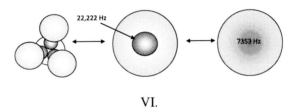

VI.

Thus, the cluster structures of vegetable oils, animal fats were concluded to be a dynamic nano disperse system in which permanently run reversible mass transfusions between different clusters but in the frame of a stable LRO with a minimal potential energy of entire mass ensemble, like below (scheme VII).

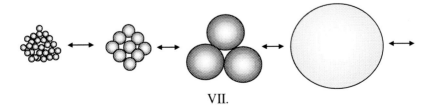

VII.

CONCLUSION

The long-range order in amorphous vegetable oils and animal fats was forcedly formed under the influence of gravitational noises, penetrating our planet.

In vegetable oils, butter and pork fat, sequences of small molecular clusters, large sub micelles and huge micelles were formed under gravitational noises.

SPONSORSHIP

For a further project developing the authors are looking for sponsors. Please contact us!

REFERENCES

[1] Zubow K., Zubow A.V., Zubow V.A. The Way to the ETIs. Applied gravitational mass spectroscopy. *Nova Sci. Publ.* NY, 2014.
[2] Belitz H.-D., Grosch W. Lehrbuch der Lebensmittelchemie. 3 Aufl. Springer-Verlag Berlin, Heidelberg NY, London, Paris, Tokyo.1987. S.144.
[3] Chaplin M., SBU London. http://www.lsbu.ac.uk/water/index.html.
[4] Horsthemke W., Lefever R. Noise-Induced Transitions. Theory and Applications in Physics, Chemistry and Biology. Edit. Hermann Haken, vol. 15. *Springer-Verlag*, Berlin, Heidelberg, NY, Tokio 1984).
[5] Frolov Y. G. Course of Colloid Chemistry. Surface phenomena and disperse systems. M., *Publ. Chimija*, 1982. P. 298.
[6] Annika Lenz; Lars Ojamäe. On the stability of dence versus cage-shaped water clusters: Quantum- chemical investigations of zero-point energies, free energies, basis-set effects and IR spectra of $(H_2O)12$ and $(H_2O)20$. *Chemical Physics Letters.* 2006, vol. 418, pp. 361-367.

In: Advances in Chemistry Research. Volume 35 ISBN: 978-1-53610-734-0
Editor: James C. Taylor   © 2017 Nova Science Publishers, Inc.

Chapter 5

CLASS INFLUENCE OF RELATIVE CONTENT OF MONTERPENES AND SESQUITERPENES AND ITS IMPACT ON THE ANTIOXIDANT ACTIVITY: APPLIED FOR SOME ALGERIAN ENDEMIC AND MEDICINAL PLANTS

Nadhir Gourine[*]
Laboratory of Research of Fundamental Sciences, University Amar Télidji of Laghouat, Laghouat, Algeria

ABSTRACT

Since the middle ages, essential oils have been widely used for bactericidal, virucidal, fungicidal, antiparasitical, insecticidal, medicinal and cosmetic applications, especially nowadays in pharmaceutical, sanitary, cosmetic, agricultural and food industries. Because of the mode of extraction, mostly by distillation from aromatic plants, they contain a variety of volatile molecules such as terpenes and terpenoids, phenol-derived aromatic components and aliphatic components. Most of the commercialized essential oils are chemotyped by gas chromatography and mass spectrometry analysis. Despite their wide use and being familiar to us as fragrances, it is important to develop a better understanding of the

---

[*] Corresponding Author e-mail: gourine.nadir@gmail.com or n.gourine@lagh-univ.dz.

influence of the two main classes of mono- and sesqui-terpenes on the antioxidant activity of their essential oils. These finding if exists, could certainly be intended for exploring new ways of applications for food additives seeking a better human health. The aim of current chapter is to **roughly investigate the possible link between the main classes' content of** monoterpenes and sesquiterpenes and to try to find a direct link between them and their antioxidant activity. This investigation was carried out using the multivariate statistical analysis tool, and it was applied for a random selection of essential oils of some common medicinal and endemic plants growing in Algeria. The main important results were: first, there is an opposite linear correlation between the percentages of oxygenated monoterpenes and those of the monoterpenes hydrocarbons. Second, and for a unique case study of *Pistacia altantica* essential oils, the antioxidant activity power measured by the scavenging of free radicals of DPPH was in good correlation with oxygenated monoterpenes percentages.

1. INTRODUCTION

Essential oils are complex mixtures of volatile compounds produced by plants. The respective main compounds are mainly derived from three biosynthetic pathways only, the mevalonate pathway leading to sesquiterpenes, the methyl-erythritol pathway leading to mono- and diterpenes, and the shikimic acid pathway *en route* to phenylpropenes. Nevertheless, there are an almost uncountable number of single substances and a tremendous variation in the composition of essential oils. Many of these volatile substances have diverse ecological functions. They can act as internal messengers, as defensive substances against herbivores, or as volatiles not only directing natural enemies to these herbivores but also attracting pollinating insects to their host. The composition of the essential oil often changes between different plant parts. Phytochemical polymorphism is often the case between different plant organs (Franz and Novak 2016).

It is a long known fact that qualitative and quantitative composition of genuine essential oils is not a standard one. In consequence of this, they possess different quality, value, and price on the market. In several cases, the real sources of variability are hard to determine. Variability in the composition of essential oils was most frequently discussed at the level of plant species and has the highest relevance from practical points of view (Németh-Zámboriné 2016).

Terpenoids are main constituents of plant-derived essential oils. Because of their pleasant odor, they are widely used in the food, fragrance, and pharmaceutical industry. Furthermore, in traditional medicine, terpenoids are also well known for their anti-inflammatory, antibacterial, antifungal, antitumor, and sedative activities. Although large amounts are used in the industry, the knowledge about their biotransformation in humans is still scarce. Yet, metabolism of terpenoids can lead to the formation of new biotransformation products with unique structures and often different flavor and biological activities compared to the parent compounds (Jäger and Höferl 2016).

Antioxidants degrade free radicals and protect the body from oxidation. There are three different types of defense mechanism, to which antioxidative enzymes, low-molecular nonenzymatic antioxidants, and repair mechanisms belong. CAT, SOD, glutathione peroxidase, and peroxidase belong to antioxidative enzymes, while nonenzymatic antioxidants include ascorbic acid, tocopherols, glutathione, ubiquinone, and β-carotene. Antioxidant activity is defined as a property of antioxidants to neutralize any free radicals. Generally speaking of antioxidants, their activity is required to act as a hydrogen donor. On the other hand, the aromatic ring plays a significant role and especially phenolic hydroxyl groups enhance the inhibition of oxidation. Antioxidant activity is also a measure for substance ability to prevent free radical concentration increment, oxidative stress, and risk for development-related diseases. And for the purpose of measuring antioxidant activity, many assays are applied such as 2,2-diphenyl-1-picrylhydrazyl (DPPH) assay, 2,2′-azino-bis(3-ethylbenzthiazoline-6)-sulfonic acid (ABTS) assay, β-carotene bleaching test, ferric and cupric reducing power, and linoleic acid assay. Many studies assess the antioxidant activity of the essential oil to be used as a natural source of antioxidants. Generally, plant species are rich in phenolic compounds, flavonoids, tannins, lignans, etc., which are all able to protect human health. Single compounds of essential oils donate protons to highly reactive radicals, inactivate it, and prevent possible damage. Various studies propagate essential oils as a great source of antioxidants and an ideal substitution of synthetic ones, such as butylatedhydroxytoluene and butylatedhydroxyanisole (BHA), which are defined as highly potent antioxidants but also with high carcinogenicity and other toxic properties as side effects (Buchbauer and Erkic 2016).

When studying essential oils, some questions might rises, such as: Is there any direct correlations in compositions between the different classes involved in the process of the biosynthesis of the compounds constituting the essential oils, and which are produced by plants. By different classes of essential oils, we meant here the four major classes and sub-classes: monoterpenes and sesquiterpenes compounds (non-oxygenated "= hydrocarbons" and oxygenated). In the same context, another major question might also arise: Are antioxidant activities related (or at least roughly related) to some specific classe of the essential oil; the answer of this question is certainly very debatable, since it is difficult to verify experimentally and practically this theory. In fact, there are thousands of plants essential oils (involving different parts to study: leaves, fruits, areal parts, roots, stems, etc., and exhibiting larges variations and fluctuations in both their relative classes' compositions, and not mentioning the exact nature of their chemical compounds). Another difficulty is the choice of the antioxidant activity assay, since there are several kinds of protocols employed that are not mandatory correlated between each others. Theses protocols are different in many ways, for example they differ in their: basic mechanisms of reaction (radical, oxidation-reduction ...), pH medium, organic or aqueous environments, reaction time (end of reaction time), solvents used (involving solubility of the reactants), and sometimes by adopting varied expressions of result referring to the same assay, etc.

The aim of the present chapter is splitted into two goals. The first objective is to investigate any possible relations or correlations of the chemical compositions among the present classes of the essential oils, obtained by random selection of some plants growing in Algeria. The second objective involves the case study of the possible link between the free radicals scavenging capacity (measured by DPPH assay) and the different classes' content of the essential oils of the leaves of *Pistacia atlantica* (Desf.) growing in Algeria.

2. MATERIALS AND METHODS

2.1. Brief Description of some Selected Algerian Plants

Artemisia herba-alba (Asso.) = [*Artemisia aragonensis* (Lam.), *Seriphidium herba-alba* (Asso) Soják], commonly known as white wormwood or desert wormwood (Arabic name chih), is a greyish-strongly aromatic dwarf shrub native to the South western Europe, Northern Africa, Arabian Peninsula and Western Asia (Belhattab et al. 2014). This plant is used for the treatment of diabetes, for its antihyperglycemic and hypoglycemic effect (Hamza et al. 2015).

Cotula cinerea (Del.) The species This plant is usually known as "Guertofa" among the local people in Northern Sahara and is commonly used in traditional medicine in the southwest of Algeria, for the treatment of several diseases like colic, cough, diarrhea, headache, and digestive disorders. All parts of the plant are used in different forms (maceration, decoction, infusion or inhalation), according to the treated diseases *Cotula cinerea*, with synonym of *Brocchia cinerea*, is a small woolly whitish plant (Djellouli et al. 2015).

Myrtus communis (L.), commonly named Myrtle, is an aromatic shrub of the Myrtaceae family, widespread all around the Mediterranean basin. In Algeria, it grows wild throughout the Tell Atlas and the coastal regions of Algiers and Constantine, where it is known as «rihan» or «mersin». *M. communis* has a long history of use as food preservative and in traditional medicine. In Algeria, the leaves of *M. communis* are used traditionally in the treatment of respiratory disorders, bronchitis, sinusitis, otitis, diarrhea and hemorrhoids (Bouzabata et al. 2015).

Pelargonium graveolens (L.) is native of South Africa and was introduced into North Africa as an ornamental plant. It has adapted well to the Mediterranean climate. Rose-scented geranium (*Pelargonium* species) is a multi-harvest high value, aromatic plant cultivated for its essential oil which is widely used in cosmetic industry and as flavouring for foods. Geranium essential oil is obtained from the scented leaves of a number of *Pelargonium* cultivars grown mainly in Reunion, China and Algeria. The volatile oil of rose geranium species has been attributed a number of biological properties including: antibacterial, antifungal and also some other pharmacological properties such as anti-inflammatory, spasmolytic and hypoglycaemic effects (Boukhatem, Kameli, and Saidi 2013).

Mentha pulegium L. (MP) (Pennyroyal) known in Algeria as 'Fliou', is one of the most frequently used herb. The essential oils and aerial parts of the plant are widely used in traditional medicine mainly for the treatment of various digestive tract diseases such as flatulence, dyspepsia and intestinal colic. Besides, the plant is used in gastronomy as culinary herb, in fragrance and pharmaceutical industries (Brahmi et al. 2016).

Mentha rotundifolia (L.) Huds **(MR) commonly known as 'apple mint' is a wild growing perennial, herbaceous plant. It is widely distributed in North Algeria in sub-humid areas, along rivers in plains and mountains where it is known as "Timija or Timarssat."** Although *Mentha suaveolens* (Ehrh.) is the synonym of *Mentha rotundifolia* (L.) Huds, the latter is nowadays known as a hybrid of *M. longifolia* and *M. suaveolens*. In Algeria *M. rotundifolia* is widely used e.g., leaf decoction is made for topical application to treat furunculosis and abscesses, to reduce fever and as a mouthwash for dental pains. In addition, the plant is reported to treat bronchitis, cough, and ulcerative colitis. It is also taken as tonic, used as stimulant, stomachic, carminative, analgesic, choleritic, antispasmodic, sedative, and hypotensive as well as a common spice (Brahmi et al. 2016).

Pinus halepensis (Mill.) belongs to the family of Pinaceae. It spreads around the Mediterranean basin including islands. Its continental range extends from Northern Africa (Morocco, Algeria, Tunisia and Libya) and Middle East (Syria, Lebanon, Jordan, Palestina and Turkey), up to Southern Mediterranean Europe (Eastern Greece, Croatia, Northern Italy, Eastern France and Eastern Spain). It can grow on poor and highly calcareous soils. *P. halepensis* is known by their medicinal properties as well as its economic importance (Djerrad, Kadik, and Djouahri 2015).

Tetraclinis articulata (Vahl.) Masters, also known as *Thuja articulata* (Vahl.) or *Callitris quadrivalvis* (Vent.) belongs to the family of Cupressaceae It is an endemic species of North Africa, Malt and Spain and constitutes a significant element of vegetation particularly in Morocco, Algeria and Tunisia. Various parts of this tree are used in folk medicine for its multiple therapeutic effects, it is mainly used against childhood, respiratory and intestinal infections, gastric pains, diabetes, hypertension, antidiarrheal, antipyretic, diuretic, antirheumatic and oral hypoglycemic (Djouahri, Boudarene, and Meklati 2013).

Eucalyptus globulus belongs to the family of Myrtaceae which is indigenous to Australia. It was introduced to Algeria in 1854 by Ramel, where it is now widely distributed. The essential oils of *Eucalyptus* species are widely used in the world, the United States Food and Drug Authority considered them as safe and non-toxic, even the Council of Europe has approved the use of *Eucalyptus* oils as flavoring agent in foods. Consequently, a growing interest has been given to their use in the scientific research field and industry as a natural food additive, drugs and cosmetics (Harkat-Madouri et al. 2015).

2.2. Source of Essential Oil Terpenes Classes Compositions (Data Sets)

For the purpose of investigation of correlations between monoterpenes (oxygenated and non-oxygenated) and sesquiterpenes (oxygenated and non-oxygenated) percentages in essential oils of a random selection of medicinal and endemic plants growing in Algeria, the data set were acquired from literature (Table 1).

Table 1. Monoterpenes and sesquiterpenes compositions for some selected Algerian medicinal plants

Plant name	N°	Total monoterpene hydrocarbons	Total oxygenated monoterpenes	Total sesquiterpene hydrocarbons	Total oxygenated sesquiterpenes	Reference
Mentha pulegium (L.)	1	1.7	92.3	1.7	1.9	(Brahmi et al. 2016)
Mentha rotundifolia (L.)	2	6.92	49.78	12.88	5.68	
Eucalyptus globulus	3	6.51	78.58	0.34	13.39	(Harkat-Madouri et al. 2015)
Artemisia herba-alba (Asso.)	4	14.2	73.1	3.2	0.1	(Belhattab et al. 2014)
	5	9.6	79.7	1.1	0.4	
	6	9.9	76.1	5.1	0.3	
	7	5.3	72.4	10.2	1.2	
Cotula cinerea (Del.)	8	2.17	95.4	0.41	0.68	(Djellouli et al. 2015)
Myrtus communis (L.)	9	57.6	31.0	0.7	0.2	(Bouzabata et al. 2015)
	10	40.4	46.3	0.8	0.5	
Ammoides verticillata (Desf.)	11	39.54	59.37	tr	tr	(Benbelaïd et al. 2014)
Artemisia arborescens (Vaill.)	12	7.93	59.05	27.24	tr	
Dittrichia graveolens (L.)	13	9.38	76.93	5.53	1.56	
Lavandula dentata (L.)	14	21.17	72.16	0.58	0.81	
Lavandula multifida (L.)	15	5.8	62.07	24.86	2.86	
Mentha piperita (L.)	16	1.79	91.09	tr	tr	
Origanum vulgare subsp. glandulosum (Desf.)	17	52.18	44.99	1.79	0.18	
Rosmarinus eriocalyx (Jord. and Fourr.)	18	33.7	56.76	0.56	2.61	
Thymbra capitata (L.) Cav.	19	22.88	67.28	1.7	tr	

Table 1. (Continued)

Plant name	N°	Total monoterpene hydrocarbons	Total oxygenated monoterpenes	Total sesquiterpene hydrocarbons	Total oxygenated sesquiterpenes	Reference
Pelargonium graveolens (L.)	20	0.2	78.9	13.9	1.3	(Boukhatem, Kameli, and Saidi 2013)
Pinus halepensis (Mill.)	21	51	28.6	0.9	13.6	(Djerrad, Kadik, and Djouahri 2015)
	22	52	29.3	1.1	14.1	
	23	52	29.4	3.4	12.4	
	24	48.1	30.2	1.8	13.1	
	25	52	29.1	2.4	12	
	26	61.3	24.0	7.2	0.8	
	27	58.7	27.3	5.6	2.8	
	28	58.1	32.1	6.0	0.8	
	29	51.3	35.3	6.0	2.9	
	30	57.6	33.0	1.6	2.1	
Tetraclinis articulate (Vahl.)	31	33.33	29.62	25.92	11.11	(Djouahri, Boudarene, and Meklati 2013)

In order to investigate the relation between antioxidant activity and the contents of different classes of terpenoids, we have selected the essential oils of the leaves of *Pistacia atlantica* (Desf.) from Algeria, for the following reasons: the availability of a large number of studied samples for both their terpenoids contents classes and their antioxidant activity measured by DPPH assay, which by the way is the wide used protocol in most published research papers. The second reason is that the samples offered larges variations in terpenoids class's percentages which are essential for this study. The data set were also acquired from literature (Gourine, Yousfi, Bombarda, Nadjemi, Stocker, et al. 2010, Gourine, Yousfi, Bombarda, Nadjemi, and Gaydou 2010).

2.3. Multivariate Statistical Analyses

Multivariate statistical analysis were investigated employing both methods of Principal component analysis (PCA) and Hierarchical Ascendant Classification (HAC) "*cluster analysis*"; and were performed using XLSTAT 2015.4.01 program.

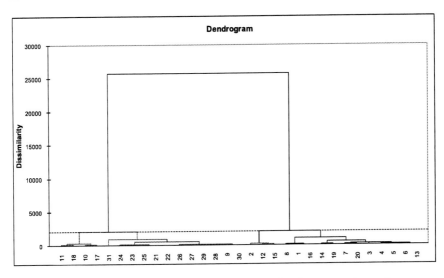

Figure 1. Dendrogram obtained from the cluster analysis of mono- and sesqui-terpenes fractions compositions employing 31 individuals of *Pistacia atlantica* (Desf.). Samples were clustered using Ward's technique with an Euclidean distance measure.

3. RESULTS AND DISCUSSION

The investigation of the relations between the main classes of terpenoids (of the essential oils monoterpenes hydrocarbons MH, oxygenated monoterpenes OM, sesquiterpenes hydrocarbons SH, oxygenated sesquiterpenes OS) of the studied plants were performed using two methods of statistical analysis: Hierarchical Ascendant Classification HAC and Principal component analysis PCA. For the HAC a data set of 31 individuals (Table 1) were subjected to cluster analysis using Ward's technique. The studied variables were M, OM, S, OS respectively. The result of the HAC was schemed in the dendrogram presented in Figure 1. This dendogram shows clearly the presence of two different groups. In addition the distance measuring the dissimilarity was very important between two separated clusters, which mean that two groups could be easily distinguished. This result was further checked with PCA. Similarly to HAC, the PCA analysis used the same individuals and variables as this previous analysis method. For the PCA method, loading factors for principal axes F1 and F2 with rotation Varimax showed strong correlations with most of considered variables (i.e., SH, OM, MH). Axis F1, which represents 54.10% of the total information, was highly positively correlated with OM ($R^2 = 0.95$), and highly negatively correlated with MH ($R^2 = -0.93$); whereas OS was in good negatively correlation with this axis ($R^2 = -0.61$). The F2 Axis, which represents 26.63% of the total information, was positively and highly correlated with only one variable which is SH ($R^2 = 0.96$). Based on theses correlations, the examination of the plots of two dimensional on axis F1 and F2 (80.73%) shows that the two clusters which were earlier separated by HAC method are separated here upon the difference in their contents of OM and MH. The right group is distinguished by highs percentages of OM and lowers percentages of MH, inversely the left group is characterized by highs percentages of MH and lowers percentages of OM. The two axis plot shows that the selected individuals are covering a large domain of variation of the four studied classes, and therefore these selected individuals will offer statically some reliable results for the correlation **investigation regarding the different classes' distribution** and also their relation with the antioxidant activity, if exists.

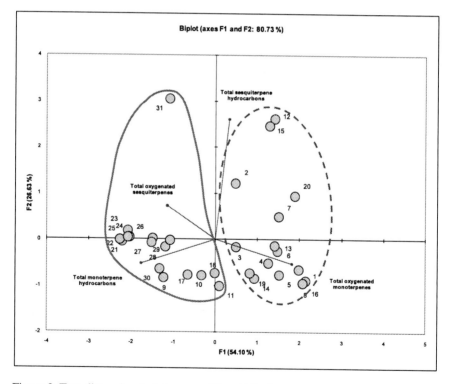

Figure 2. Two dimensional plot on axes F1 and F2 of mono- and sesqui-terpenes fractions compositions of individuals of *Pistacia atlantica* (Desf.) using Principal Component Analysis PCA.

The next logical step is to plot different classes versus each other and try to find some linear correlations between them; this work was done, and the classes presenting good to excellent correlation were presented in figures 3-a to 3-d. The figure 3-a shows an excellent linear correlation between OM and all oxygenated compounds ($R^2 = 0.954$). Furthermore, figure 3-b present a good correlation of MH with total non oxygenated compounds ($R^2 = 0.883$). In another hand, an acceptable correlation ($R^2 = 0.691$) was noticed between SH and total sesquiterpenes compounds (Fig. 3-c). Finally, a good correlation ($R^2 = 0.819$) with a negative slope was established between MH and OM which was demonstrated earlier by the presence of the two groups 'clusters' by HAC and PCA methods.

The second part of this current work is the determination of the possible relation of the antioxidant activity (measured by DPPH assay and expressed as $1/IC_{50}$) with one of the four different terpenoid classes' (MH, OM, SH, OS). So, in order to achieve this task, the PCA method was employed. The data set was composed of 43 samples and five variables (MH, OM, SH, OS, $1/IC_{50}$). The loading factors for principal axes F1 and F2 with rotation Varimax and total axis information of 68.49%, showed strong correlations only with two variables (i.e., OM and OS) (figure 4). The axis F1, which represents 36.59% of the total information, was highly positively correlated with OS ($R^2 = 0.87$), and in good correlated with SH ($R^2 = 0.72$); whereas MH was in moderate negatively correlation with this axis ($R^2 = -0.69$). The F2 Axis, which represents 31.90% of the total information, was positively and highly correlated with only one variable which is OM ($R^2 = 0.83$), but present moderate correlations positively with $1/IC_{50}$ ($R^2 = 0.59$), and negatively with M ($R^2 = -0.63$). A part what is was said, the main important result is that OM are highly correlated with $1/IC_{50}$. As a conclusion the antioxidant capacity of essential oil is correlated to the content of oxygenated monoterpenes in the essential oil of the leaves of *Pistacia atlantica*.

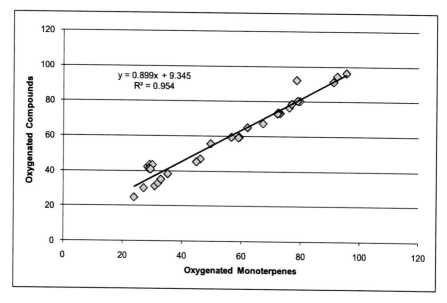

Figure 3-a. Linear regression relation obtained between compositions of oxygenated monoterpenes and oxygenated compounds for a selected set of the essential oils from medicinal plants growing in Algeria.

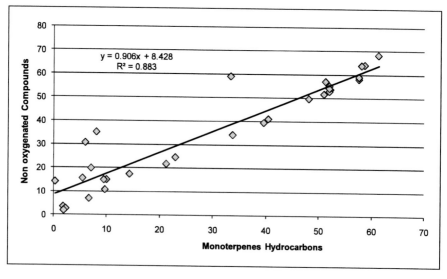

Figure 3-b. Linear regression relation obtained between compositions of monoterpenes hydrocarbons and non oxygenated compounds for a selected set of the essential oils from medicinal plants growing in Algeria.

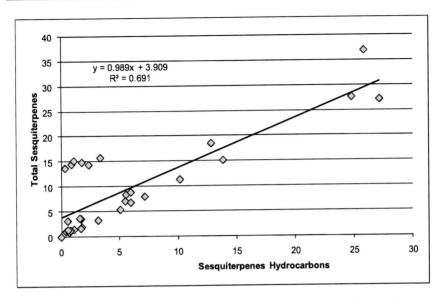

Figure 3-c. Linear regression relation obtained between compositions of sesquiterpenes hydrocarbons and total sesquiterpenes for a selected set of the essential oils from medicinal plants growing in Algeria.

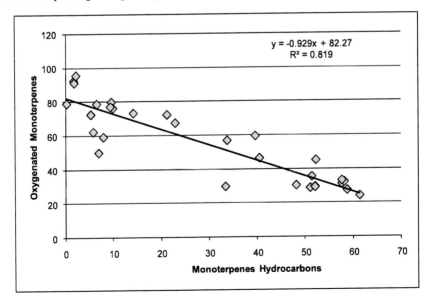

Figure 3-d. Linear regression relation obtained between compositions of monoterpenes hydrocarbons and oxygenated monoterpenes for a selected set of the essential oils from medicinal plants growing in Algeria.

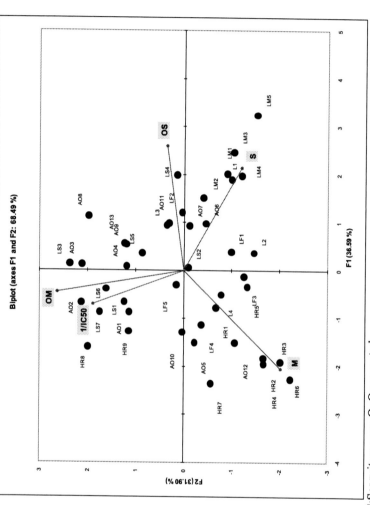

M: Monoterpenes; S:Sesquiterpenes; O: Oxygenated.

Figure 4. Two dimensional plot on axes F1 and F2 of antioxidant power (1/IC$_{50}$) and mono- and sesqui-terpenes fractions compositions of individuals of the leaves essential oils of *Pistacia atlantica* (Desf.) using Principal Component Analysis PCA.

CONCLUSION

The existence of different classes of terpenoids constituting the essential oil extracts that are biosynthesized by plants according to a certain pathway relative to each type of plant and its environment leads to ask some questions about the relative variations in the contents of these terpenoid classes; and whether or not the percentages distribution of these classes are following some kind of rule or not. By this present investigation, we may say that there are in fact some correlations between some classes of the terpenoids constituting the essential oils of the plants. Furthermore, for a case study, we have found here in this investigation that the antioxidant activity of the essential oil of *Pistacia atlantica* measured by DPPH assay is in good correlation with oxygenated monoterpenes class; this humble result opens the sights for more profound investigations about the correlation between the antioxidant activity measured by other techniques and the terpenoid classes for different essential oils.

REFERENCES

Belhattab, Rachid, Loubna Amor, José G. Barroso, Luis G. Pedro, and A. Cristina Figueiredo. 2014. "Essential oil from *Artemisia herba-alba* Asso grown wild in Algeria: Variability assessment and comparison with an updated literature survey." *Arabian Journal of Chemistry* no. 7 (2):243-251. doi: http://dx.doi.org/10.1016/j.arabjc.2012.04.042.

Benbelaïd, Fethi, Abdelmounaïm Khadir, Mohamed Amine Abdoune, Mourad Bendahou, Alain Muselli, and Jean Costa. 2014. "Antimicrobial activity of some essential oils against oral multidrug–resistant Enterococcus faecalis in both planktonic and biofilm state." *Asian Pacific Journal of Tropical Biomedicine* no. 4 (6):463-472. doi: http://dx.doi.org/10.12980/APJTB.4.2014C1203.

Boukhatem, Mohamed Nadjib, Abdelkrim Kameli, and Fairouz Saidi. 2013. "Essential oil of Algerian rose-scented geranium (*Pelargonium graveolens*): Chemical composition and antimicrobial activity against food spoilage pathogens." *Food Control* no. 34 (1):208-213. doi: http://dx.doi.org/10.1016/j.foodcont.2013.03.045.

Bouzabata, Amel, Célia Cabral, Maria José Gonçalves, Maria Teresa Cruz, Ange Bighelli, Carlos Cavaleiro, Joseph Casanova, Félix Tomi, and Ligia Salgueiro. 2015. "*Myrtus communis* L. as source of a bioactive and safe essential oil." *Food and Chemical Toxicology* no. 75:166-172. doi: http://dx.doi.org/10.1016/j.fct.2014.11.009.

Brahmi, Fatiha, Adjaoud Abdenour, Marongiu Bruno, Porcedda Silvia, Piras Alessandra, Falconieri Danilo, Yalaoui-Guellal Drifa, Elsebai Mahmoud Fahmi, Madani Khodir, and Chibane Mohamed. 2016. "Chemical composition and in vitro antimicrobial, insecticidal and antioxidant activities of the essential oils of *Mentha pulegium* L. and *Mentha rotundifolia* (L.) Huds growing in Algeria." *Industrial Crops and Products* no. 88:96-105. doi: http://dx.doi.org/10.1016/j.indcrop.2016.03.002.

Buchbauer, Gerhard, and Marina Erkic. 2016. "Antioxidative Properties of Essential Oils and Single Fragrance Compounds." In: *Handbook of essential oils: science, technology, and applications*, edited by K Husnu Can Baser and Gerhard Buchbauer, 323-344. New York: CRC Press.

Djellouli, Mohammed, Houcine Benmehdi, Siham Mammeri, Abdellah Moussaoui, Laid Ziane, and Noureddine Hamidi. 2015. "Chemical constituents in the essential oil of the endemic plant *Cotula cinerea* (Del.) from the southwest of Algeria." *Asian Pacific Journal of Tropical Biomedicine* no. 5 (10):870-873. doi: http://dx.doi.org/10.1016/j.apjtb.2015.06.007.

Djerrad, Zineb, Leila Kadik, and Abderrahmane Djouahri. 2015. "Chemical variability and antioxidant activities among *Pinus halepensis* Mill. essential oils provenances, depending on geographic variation and environmental conditions." *Industrial Crops and Products* no. 74:440-449. doi: http://dx.doi.org/10.1016/j.indcrop.2015.05.049.

Djouahri, Abderrahmane, Lynda Boudarene, and Brahim Youcef Meklati. 2013. "Effect of extraction method on chemical composition, antioxidant and anti-inflammatory activities of essential oil from the leaves of Algerian *Tetraclinis articulata* (Vahl) Masters." *Industrial Crops and Products* no. 44:32-36. doi: http://dx.doi.org/10.1016/j.indcrop.2012.10.021.

Franz, Chlodwig, and Johannes Novak. 2016. "Sources of Essential Oils." In: *Handbook of essential oils: science, technology, and applications*, edited by K Husnu Can Baser and Gerhard Buchbauer, 43-86. New York: CRC Press.

Gourine, N., M. Yousfi, I. Bombarda, B. Nadjemi, P. Stocker, and E. M. Gaydou. 2010. "Antioxidant activities and chemical composition of essential oil of *Pistacia atlantica* from Algeria." *Industrial Crops and Products* no. 31 (2):203-208. doi: 10.1016/j.indcrop.2009.10.003.

Gourine, Nadhir, Mohamed Yousfi, Isabelle Bombarda, Boubakeur Nadjemi, and Emile Gaydou. 2010. "Seasonal Variation of Chemical Composition and Antioxidant Activity of Essential Oil from *Pistacia atlantica* Desf. Leaves." *Journal of the American Oil Chemists Society* no. 87 (2):157-166. doi: 10.1007/s11746-009-1481-5.

Hamza, Nawel, Bénédicte Berke, Catherine Cheze, Sébastien Marais, Simon Lorrain, Abdelilah Abdouelfath, Regis Lassalle, Dominique Carles, Henri Gin, and Nicholas Moore. 2015. "Effect of *Centaurium erythraea* Rafn, *Artemisia herba-alba* Asso and *Trigonella foenum-graecum* L. on liver fat accumulation in C57BL/6J mice with high-fat diet-induced type 2 diabetes." *Journal of Ethnopharmacology* no. 171:4-11. doi: http://dx.doi.org/10.1016/j.jep.2015.05.027.

Harkat-Madouri, Lila, Boudria Asma, Khodir Madani, Zakia Bey-Ould Si Said, Peggy Rigou, Daniel Grenier, Hanane Allalou, Hocine Remini, Abdennour Adjaoud, and Lila Boulekbache-Makhlouf. 2015. "Chemical composition, antibacterial and antioxidant activities of essential oil of *Eucalyptus globulus* from Algeria." *Industrial Crops and Products* no. 78:148-153. doi: http://dx.doi.org/10.1016/j.indcrop.2015.10.015.

Jäger, Walter, and Martina Höferl. 2016. "Metabolism of Terpenoids in Animal Models and Humans." In: *Handbook of essential oils: science, technology, and applications*, edited by K Husnu Can Baser and Gerhard Buchbauer, 253-279. New York: CRC Press.

Németh-Zámboriné, Éva 2016. "Natural Variability of Essential Oil Components." In: *Handbook of essential oils: science, technology, and applications*, edited by K Husnu Can Baser and Gerhard Buchbauer, 87-125. New York: CRC Press.

Chapter 6

CHEMICAL COMPOSITION AND BIOLOGICAL ACTIVITIES OF ESSENTIAL OILS FROM THREE MEDICINAL PLANTS OF NIGERIA

Oladipupo A. Lawal[1,*], Isiaka A. Ogunwande[1,*], Bolanle T. Jinadu[1], Omolara T. Bejide[2] and Emmanuel E. Essien[3]

[1]Natural Products Research Unit, Department of Chemistry, Faculty of Science, Lagos State University, Ojo, Lagos, Nigeria
[2]Department of Chemical Sciences, Faculty of Science, Olabisi Onabanjo University, Ago-Iwoye, Ogun State, Nigeria
[3]Department of Chemistry, Faculty of Science, University of Uyo, Akwa-Ibom State, Nigeria

ABSTRACT

The chemical compositions and biological activities of essential oils hydrodistilled from three plants from Nigeria are being reported. The constituents of the essential oils were analysed by means of gas chromatography (GC) and gas chromatography-mass spectrometry (GC-

---

[*] Corresponding Author Address: Natural Products Research Unit, Department of Chemistry, Faculty of Science, Lagos State University, PMB 0001 LASU Post Office, Ojo, Lagos, Nigeria. E-mail:jumobi.lawal@lasu.edu.ng; isiaka.ogunwande@lasu.edu.ng.

MS) techniques. The antibacterial, insecticidal, larvicidal and phytotoxicity studies were evaluated by standard procedures. The major components of the leaf oil of *Blighia sapida* K.D. Koenig were 6,10,14-trimethyl-2-pentadecanone (12.8%), geranyl acetone (12.0%), phytol (10.8%) and α-ionone (6.1%). The main constituents of the flower oil of *Thevetia peruviana* Pers (Apocynaceae) were β-ionone (44.5%), terpinen-4-ol (8.9%) and, terpinolene (8.3%). However, limonene (30.0%), δ-cadinene (13.3%), α-copaene (10.1%) and terpinolene (10.0%) were the main compounds of the leaf oil. Moreover, 1,8-cineole (54.1%) and α-terpineol (15.6%) represent the main compounds of the seeds of *Aframomum longiscapum* (Hook.f.) K. Schum while the pod consists mainly of linalool (34.3%), 1,8-cineole (15.7%) and β-pinene (11.0%). In addition, β-pinene (27.9%), β-caryophyllene (18.8%), caryophyllene oxide (12.2%) and α-pinene (9.7%) were the constituents occurring in higher proportions in the leaf.

The *B. sapida* oil exhibited strongest antibacterial activity against *Staphylococcus aurues* and *Escherichia coli* with the minimum inhibitory concentration (MIC) of 0.3 mg/mL, *Bacillus subtilis* (MIC, 0.6 mg/mL), *Pseudomonas* spp. (MIC, 1.2 mg/mL) and *Kliebsiella* spp. (MIC, 1.3 mg/mL). The studied essential oil of *A. longisparcum* oil displayed strong antibacterial activity against *B. subtilis* (MIC: leaf and pod, 0.6 mg/mL; seed, 0.3 mg/mL), *S. aurues* (MIC: leaf, 0.6 mg/mL; seed, 0.1 mg/mL; pod, 0.3 mg/mL), *E. coli* (MIC: leaf, 1.2 mg/mL; seed and pod, 0.3 mg/mL), *Kliebsiella* spp. (MIC: seed, 0.6 mg/mL; pod, 1.2 mg/mL), *Pseudomonas* spp. (MIC: leaf, seed and pod, 1.2 mg/mL) and *Proteus* spp. (MIC: seed, 1.2 mg/mL).

The essential oils of *B. sapida* and *T. peruviana* displayed strong insecticidal activity against *Sitophilus zeamais*. The lethal concentrations (LC_{50}) were 6.28 mg/L air (*B. sapida* leaf), 1.52 mg/L (*T. peruviana* flower) and 4.35 mg/L air (*T. peruviana* leaf). The result of larvicidal activity of *B. sapida* leaf oil against the fourth-in-star larvae of *Anopheles gambiae* revealed a lethal concentration (LC_{50}) of 11.61 mg/L air. The phytotoxicity activity *T. peruviana* essential oils on seeds germination and seedling growth of *Zea mays* indicated the oils to have percentage germination ranging from 76.6% to 100% against the seeds of *Zea mays* at 25 to 500 mg/mL in dose dependent manner.

Keywords: *Blighia sapida*, *Thevetia peruviana*, *Aframomum longiscapum*, essential oil compositions, terpenes, biological activities

INTRODUCTION

In continuation of our study on the terpenes components and biological activity of Nigerian medicinal plants and herbs [1], we report the chemical constituents, antibacterial, insecticidal, larvicidal and phytotoxicity potentials of essential oils from three plants. *Blighia sapida* (known *'ackee'* and in Nigeria as *'isin'*) is an evergreen tree, native to West African wild forests. The fruits of *B. sapida* may be eaten raw (without the pink raphe attaching the aril to the seed) or after cooking when it resembles scrambled eggs. The aril is eaten fresh while the seed can be exploited for human food. Extracts of the plant have exhibited molluscicidal activity against *Bulinus globosus* and *Bulinus truncate* [2], displayed larvicidal activity against 4th-instar larvae of *Aedes aegypti* [3], mosquito larvae of *Anopheles gambiae* and *Culex quinquefasciatus* [4] and exhibited pesticidal activity against the insect pests, *Tribolium castaneum*, *Callosobruchus maculatus* and *Sitophilus zeamais* [5]. In addition extracts of ackee have shown potential inhibition of α-amylase and α-glucosidase enzymes [6] and possess antioxidant activity [7, 8]. Moreover, the antimicrobial [9, 10] and prophylactic potentials of polypenols [11] from the plant have been evaluated. The seed contain nutrients such as protein and carbohydrate [12], fatty acids [13, 14] and amino acids [15]. Phytochemicals isolated from ackee include hypoglycin A and B, which have shown hypoglycemia in animals [16], methyl gallate, isoquercitrin, protocatechuic acid, gallic acid, ellagic acid and quercetin [7], blighoside A-C, stigmasta-5,22-dien-3-ol and stigmasta-5,22-dien-3-*O*-glucopyranoside [17], blighinone [18], vomifoliol [19], saponins [20], (2S,1'S,2'S)-2-(2'-carboxycyclopropyl) glycine [21] and dodecahydro-1H-cyclopenta[a] phenanthren-10-ol [22].

Thevetia peruviana Pers., is an evergreen tropical shrub or small tree that bears yellow or orange-yellow, trumpet like flowers and its fruit is deep red/black in colour encasing a large seed. The plant originates from tropical America and is widely cultivated throughout the tropics as an ornamental, also in tropical Africa [23] [1]. These plants are toxic to most vertebrates as they contain cardiac glycosides. Extracts of *T. peruviana* were found to possessed moluscicidal [24, 25], rodenticidal [26], remarkable antitumor activity [27], antimicrobial [28-31], antispermatogenic [32] and antioxidant [30, 33] activities. In addition, the effect of *T. peruviana* on *Parthneium hysterophorus* [34] and allelophatic activity on the growth of *Triticum aestivum* [35] have also been reported. Extract of *T. peruviana* showed high anti HIV-1 and HIV-1 IN inhibitory activity [36]. The molluscicidal activity of apigenin from the plant was time as well as dose depend [37]. The termite controlling properties and cytotoxicity effects of *T. peruviana* were known [38]. *Thevetia peruviana* plant was also reported to exhibited piscicidal activity [39], shown toxicity to insect pest [40] and displayed larvicidal efficacy against the aquatic stages of *Aedes aegypti* [41] and *Anopheles stephensi* [42]. A study showed that *T. peruviana* latex is a rich source of pathogenesis-related proteins, including cysteine peptidases [43] and fatty acids [44]. The plant contained cardiac glycosides such as thevetin C, acetylthevetin C, thevetin B and acetylthevetin B [45, 46], 19-nor-cardenolides and 19-nor-10-hydroperoxycardenolides which are cytotoxic to some human cell cancers [46]. The anti-inflammatory activity of quercetin, kaempferol and quercetin-7-*o*-galactoside from *T. peruviana* has been documented [47]. The anti-HIV activity of pervianoside I, peruvianoside II, peruvianoside III is well known [48]. The essential oil and its components have been reported to possess antinociceptive and gastroprotective effects [49], improve sleep and reduce anxiety [49]. Volatile oil showed antiplatelet and antithrombotic activities [50]. Previous analysis [49] revealed the main constituents of the essential as 1,8-cineole (34.2%), linalool (23.0%) and *cis*-dihydrocarveol (8.9%). The volatile components in the flower of *T. peruviana* [51] were palmitic acid (22.95%), (E, β)-ionone (17.2%) and 1,8-cineole (10.6%).

Aframomum longiscapum (Hook.f.) K. Schum (Zingiberaceae) is a monocotyledonous plant species. It is an erect herb with leafy stem 5-9 ft. High with mauve flowers and dark red fruits [52]. It is mainly produced in the south-western part of Nigeria as spice. The plant contains both nutrients and anti-nutrients agents [53]. Extracts of the plant mitigated the liver gamma-radiation-induced damage probably by increasing antioxidant activities [54]. The antioxidant activities of the aqueous and methanolic extracts were attributed to the presence of gingerol in the extracts [55]. The *A. longiscapum* extract exhibited anticlastogenic and hepatoprotective potentials, reduced sperm count, motility, with no effect on viability and morphology [56] and are useful for improving penile rigidity and/or preventing erectile dysfunction, including premature ejaculation, of a male mammal [57]. A previous report [58] on the volatile constituents identified β-pinene (42.6%) and β-caryophyllene (25.4%) as the main compounds in the leaf while the rhizome contained β-pinene (40.2%) and linalool (27.4%).

MATERIALS AND METHOD

Plant Materials

Fresh leaves of *B. sapida* were collected from Igando town, Ikotun-Igando Local Government Development Area, Lagos State, Nigeria, in July 2012. On the other hand, fresh leaves and flowers of *T. peruviana* were collected from Lagos State University, Ojo campus, Lagos State, Nigeria, in May 2015. Botanical identification was carried out at the Department of Botany, University of Lagos, Akoka-Yaba, Lagos, by Mr. T. K Odewo, where voucher specimens (LUH 5745 and LUH 6301 respectively) were deposited. However, plant materials of *A. longiscapum* were purchased from markets in Lagos State, Nigeria. Identification of the plant was carried out at the Herbarium of Forestry Research Institute of Nigeria (FRIN), Ibadan, where a voucher specimen, FHI 107666, was deposited for future reference.

Hydrodistillation of Volatile Oils

Air-dried and pulverized samples were subjected to separate hydrodistillation using Clevenger-type apparatus for 3 h in accordance with the British Pharmacopoeia specification [59]. Briefly, 250 g of each of the pulverized samples were carefully introduced into a 5 L flask and distilled water was added until it covers the sample completely. Hydrodistillation was carried out in an all glass Clevenger-type distillation unit designed according to the specification. The volatile oils distilled over water and were collected in the receiver arm of the apparatus into a separate, clean and previously weighed sample bottles. The oils were kept under refrigeration until the moment of analyses.

Analysis of the Essential Oils

Gas chromatography (GC) analysis of the oils was carried out on a Hewlett Packard HP 6820 Gas Chromatograph equipped with a FID detector and HP-5MS column (60 m x 0.25 mm i.d., film thickness 0.25 μm) and the split ratio was 1:25. The oven temperature was programmed from 50 °C (after 2 min) to 240 °C at 5 °C/min and the final temperature was held for 10 min. Injection and detector temperatures were maintained at 200 °C and 240 °C, respectively. Hydrogen was the carrier gas (1 mL/min). An aliquot (0.5 µL of the diluted oil) was injected into the GC. Peaks were measured by electronic integration. A homologous series of *n*-alkanes (C_4-C_{32}) were run under the same conditions for determination of retention indices. The relative amounts of individual components were calculated based on the GC peak area (FID response) without using correction factors.

Gas chromatography-mass spectrometry (GC-MS) analysis was performed on a Hewlett Packard Gas Chromatograph HP 6890 interfaced with a Hewlett Packard 5973 Mass spectrometer system equipped with a HP-5MS capillary column (30 m x 0.25 mm i.d., film thickness 0.25 µm). The oven temperature was programmed from 70-240°C at the rate of 5°C/min. The ion source was set at 240°C and electron ionization at 70 eV. Helium was used as the carrier gas at a flow rate of 1mL/min. The scanning range was 35 to 425 amu. Diluted oils in *n*-hexane (1.0 µL) were injected into the GC/MS.

Identification of Constituents of the Oils

The identification of constituents was performed on the basis of retention indices (RI) determined by co-injection with reference to a homologous series of *n*-alkanes or authentic samples under identical experimental conditions. Further identification was performed by comparison of their mass spectra with those from NIST [60] and the home-made MS library built up from pure substances and components of known essential oils, as well as by comparison of their retention indices with literature values as previously reported [1].

Biological Assays

Antibacterial Assay

This assay was tested against eight local bacterial isolates (two Gram-positive and six Gram-negative strains) obtained from the Nigeria Institute of Medical Research (NIMR), Yaba, Lagos and Department of Microbiology, Lagos State University, Ojo, Lagos, Nigeria. These microbes were *Bacillus subtilis*, *Staphylococcus aureus*, *Citrobacter youagae*, *Escherichia coli*, *Kiebsiella* spp., *Proteus* spp., *Pseudomonas* spp. and *Salmonella* spp. The stock cultures were maintained at 4°C in Müeller-Hinton agar (Oxoid, Germany).

Agar Disc Diffusion

The agar disc diffusion method was employed for the determination of antibacterial activity of the essential oil *B. sapida* [61]. The microorganisms were grown overnight at 37°C in 20 mL of Müeller-Hinton broth (MHB). The cultures were adjusted with sterile saline solution to obtain turbidity comparable to that of McFarland no. 5 standard (1.0 x 10^8 CFU/mL). Ninety millimeters of Petri dishes containing 12 mL of sterilized Müeller-Hinton agar were inoculated with the microbial suspensions. Sterile Whatman No.1 (6 mm) discs papers were individually placed on the surface of the seeded agar plates and 10 µL of the oil in dimethylsulfoxide (DMSO) was applied to the filter paper disk. The plates were incubated at 37°C for 24 h and the diameter of the resulting zones of inhibition (IZ) was measured. All tests were performed in triplicates. Gentamycin and tetracycline were used as positive controls, while, hexane and DMSO served as negative controls.

Minimum Inhibitory Concentration

The minimum inhibitory concentration (MIC) of *B. sapida* oil was determined using 96-well microtitre dilution method [62]. Bacterial cultures were incubated in Müller-Hinton broth overnight at 37°C and a 1:1 dilution of each culture in fresh MHB was prepared prior to use in the micro dilution assay. A two-fold serial doubling dilution of the oil was made to obtain concentrations ranging from 10 mg/mL to 0.078 mg/mL. One hundred µL of bacterial culture of an approximate inoculum size of 1.0 x 10^8 CFU/mL was added to all well and incubated at 37°C for 24 h. After incubation, 40 µL of 0.2 mg/mL *p*-iodonitotetrazolium violet (INT) solution was added to each well and incubated at 37°C. Plates were examined after about 30-60 min of incubation. MIC is defined as the lowest concentration that produces an almost complete inhibition of visible micro-organism growth in liquid medium. Solvent controls (DMSO and hexane) and the standard antibiotics (gentamycin and tetracycline) were included in the assay.

Insecticidal Activity

The fumigant toxicity of *B. sapida* oil was assayed according to the method earlier described [63]. Adult insects of mixed sex, 7-14 days old of *Sitophilus zeamais* reared on maize at 25 ± 1°C and 65% ± 5% relative humidity (R.H.) was used for the bioassay. Filter paper (Whatman No. 1, cut into 2-cm diameter pieces) was impregnated with the oil at doses calculated to give equivalent fumigant concentrations of 10-100 mg/L air. The impregnated filter paper was then attached to the under surface of the Petri dishes (90 mm) containing 10 adults of *S. zeamais* to different concentrations of the oil. Each concentration and the control were replicated three times. Mortality was determined after 24, 48 and 72 h from the commencement of exposure. When no leg movement was observed, insects were considered dead.

Larvicidal Activity

Larvae of *Anopheles gambiae* identified by an entomologist from the Department of Zoology, Lagos State University, Ojo, were collected from the storage water tank in the ETF building, Faculty of Science, Lagos State University, Ojo and were maintained at ambient rearing conditions. All bioassays were conducted at 27 ± 1°C, 65 ± 5°C RH and 12 h light and 12 h dark photoperiod [64]. A 5% yeast suspension was used as food source.

Test for mosquito larval activity was conducted according to established procedures [65, 66] with some modifications. Ten fourth-instar mosquito larvae (*A. gambiae*) were collected and transferred to the beakers (100 mL) each containing 29.0 mL of degassed distilled water and 1000 µL of different concentrations of *B. sapida* oil (10-100 ppm) in DMSO (Dimethyl sulfoxide) solution. Each test was performed in triplicate. The control was prepared with 29.0 mL of degassed distilled water and 1000 µL of DMSO solution without the oil. Observation on larval mortality was recorded after 24, 48 and 72 h exposure, during which no food was given to the larvae. Larvae were considered dead, when they did not react to touching with a needle. The percentage of mortality and lethal concentrations (LC_{50}) values were determined using Abbots formula and probit analysis program, version 1.5 respectively. Larvicidal activity was reported as LC_{50} with 95% confidence intervals, representing the concentrations in mg/mL with 50% larvae mortality rate in 72 h.

Phytotoxicity Activity

Germination and seedling growth of the seeds of maize (*Zea mays* L.) were used to study the allelopathic effects of *T. peruviana* essential oils according to the established method [67]. Seeds of *Z. mays* used in study were purchased from Iyana-Iba market, Ojo Local Government Area, Lagos,

Nigeria, in June 2015. The seeds were surface-sterilized in 95% ethanol for 15 s and sown in Petri dishes (Ø = 90 mm), containing two layers of Whatman filter paper No.1, impregnated with 10 mL of test- solutions of each oil at the different concentrations (20-250 mg/mL), 10 mL of 1% DMSO and distilled water as controls, respectively. Each bioassay was repeated three times with 10 seeds for each determination at 27 ± 1°C with natural photoperiod. The percentage germination and (root and shoot) growth for each experiment was measured after 7 days of incubation period.

Statistical Analysis

The mean and standard deviation of three experiments were determined. Statistical analysis of the differences between mean values obtained for experimental groups were calculated using Microsoft excel program, 2003. Data were subjected to one way analysis of variance (ANOVA). P values ≤ 0.05 were regarded as significant and P values ≤ 0.01 as very significant. Mortality percentages were calculated by the correction formula for natural mortality in the untreated control [68]. The Lethal concentrations (LC$_{50}$) values for the insecticidal and larvicidal activities were calculated using probit analysis program, version 1.5.

RESULTS AND DISCUSSION

Chemical Constituents of the Essential Oils

Table 1 indicated the percentages chemical compounds identified in the studied essential oil samples grown in Nigeria, according to their elution order on HP-5MS column.

Table 1. Chemical composition of studied essential oils from Nigeria

Compounds [a]	RI [b]	RI [c]	B. sapida	T. peruviana Leaf	T. peruviana Flower	A. longispacum Leaf	A. longispacum Seed	A. longispacum Pod
α-Pinene	939	932	-	-	-	9.7	-	-
β-Pinene	981	987	-	-	-	27.9	21.7	5.0
Pseudocumene	1005	1002	-	3.0	7.7	-	-	-
o-cymene	1006	1005	-	-	6.6	-	-	-
α-Terpinene	1018	1016	-	3.6	-	-	-	-
p-Cymene	1024	1024	-	-	-	-	13.1	-
Limonene	1032	1024	1.7	-	30.0	1.4	-	3.6

Table 1. (Continued)

Compounds [a]	RI [b]	RI [c]	B. sapida	T. peruviana Leaf	T. peruviana Flower	A. longispacum Leaf	A. longispacum Seed	A. longispacum Pod
1,8-Cineole	1033	1033	-	7.1	-	1.3	6.8	54.1
α-Phellandrene	1036	1033	-	-	-	1.6	-	-
Phenyl acetaldehyde	1048	1049	2.2	-	-	-	-	-
Pinocarvone	1061	1061	-	-	-	-	0.6	-
Acetophenone	1063	1059	0.6	-	-	-	-	-
γ-Terpinene	1066	1065	-	-	-	0.3	2.0	-
cis-Linalool oxide (furanoid)	1071	1067	0.6	-	-	-	-	-
p-Cresol	1089	1071	0.9	-	-	-	-	-
Terpinolene	1093	1093	-	8.3	10.0	-	0.3	-
Linalool	1101	1095	1.2	-	-	7.9	0.4	4.0
n-Nonanal	1107	1100	3.6	-	-	-	-	-
Isophorone	1115	1118	0.4	-	-	-	-	-
Myrcenol	1118	1118	-	-	-	-	-	2.1
cis-p-Menth-2-en-1-ol	1119	1118	1.4	2.1	-	-	-	-
trans-Verbenol	1148	1144	0.9	-	-	-	-	-
trans-Pinocarveol	1151	1151	-	-	-	1.0	2.0	-
(E,Z)-2,6-Nonadienal	1156	1155	1.0	-	-	-	-	-
Pinocarvone	1162	1160	2.5	-	-	0.5	-	-
(E)-2-Nonenal	1167	1162	1.3	-	-	-	-	-
Menthol	1170	1167	0.8	-	-	-	-	-
Terpinen-4-ol	1177	1177	-	8.9	-	0.3	0.5	3.9
Naphthalene	1179	1178	2.4	-	-	-	-	-
Cryptone	1184	1184	-	-	-	-	1.0	-
α-Terpineol	1189	1187	-	-	-	-	-	15.6
Myrtenol	1198	1196	-	-	-	5.4	-	-
α-Campholene aldehyde	1199	1199	-	-	-	-	1.1	-
n-Decanal	1203	1201	1.0	-	-	-	-	-
Myrtenol	1203	1203	-	-	-	-	3.9	-
β-Cyclocitral	1217	1217	1.1	-	-	-	-	-
Citronellol	1225	1223	0.8	-	-	-	-	-
Pulegone	1240	1233	1.1	-	-	-	-	-
Neral	1242	1235	1.6	-	-	-	-	-
Carvone	1249	1249	-	-	-	-	-	1.7
Cumin aldehyde	1253	1253	-	-	-	-	0.2	-
Geranial	1270	1264	1.6	-	-	-	-	-
Isobornyl acetate	1282	1282	-	-	-	-	0.8	-
4-Ethylguaiacol	1289	1280	2.1	-	-	-	-	-
trans-Pinocarvyl acetate	1296	1296	-	-	-	0.1	6.0	-
Carvacrol	1309	1311	-	1.0	-	-	-	-
Myrtenyl acetate	1335	1335	-	-	-	1.0	10.8	-
γ-Elemene	1340	1338	-	-	-	0.3	-	-
α-Terpinyl acetate	1352	1352	-	-	-	0.1	-	3.1
Eugenol	1361	1356	1.3	-	-	-	-	-
Neryl acetate	1371	1371	-	-	-	-	0.1	-

Compounds [a]	RI [b]	RI [c]	B. sapida	T. peruviana Leaf	T. peruviana Flower	A. longispacum Leaf	A. longispacum Seed	A. longispacum Pod
α-Copaene	1376	1374	-	2.4	10.1	0.2	-	-
β-Maaliene	1378	1378	-	-	-	-	1.3	-
β-Elemene	1390	1389	-	-	-	0.3	-	-
β-Caryophyllene	1417	1419	-	-	-	18.8	2.7	-
(E)-α-Ionone	1434	1428	6.1	-	-	-	-	-
α-Gurjunene	1435	1435	-	-	-	-	0.3	-
trans-α-Bergamotene	1437	1432	2.4	-	-	-	-	-
Dihydro-β-ionone	1446	1447	-	6.1	-	-	-	-
Aromadendrene	1453	1451	-	1.5	5.1	-	-	-
α-Humulene	1454	1452	-	-	-	2.8	1.1	-
(E)-β-Farnesene	1454	1453	-	-	-	0.1	-	-
Geranyl acetone	1454	1453	12.0	-	-	-	-	-
γ-Gurjunene	1470	1471	-	-	-	-	0.1	-
Germacrene D	1485	1484	-	-	-	1.0	-	-
(E)-β-Ionone	1486	1487	2.2	44.5	-	-	-	-
β-Selinene	1489	1489	-	-	-	-	0.2	-
α-Elemene	1490	1493	-	-	-	0.4	-	-
(Z,E)-α-Farnesene	1490	1495	2.2	-	-	-	-	-
β-Bisabolene	1510	1505	3.3	-	-	0.2	-	-
2,4-Di-tert-butylphenol	1518	1518	-	6.1	8.5	-	-	-
δ-Cadinene	1526	1522	-	3.3	13.3	-	-	-
Elemol	1548	1548	-	-	-	1.3	-	-
(E)-Nerolidol	1558	1561	0.9	-	-	0.8	2.0	-
Germacrene B	1561	1557	-	0.3	6.0	-	-	-
Caryophyllene oxide	1586	1586	-	-	-	12.2	12.0	-
β-Eudesmol	1648	1646	-	-	-	0.6	-	-
Patchouli alcohol	1653	1656	1.4	-	-	-	-	-
α-Bisabolol	1685	1683	-	-	-	0.4	-	-
Juniper camphor	1688	1688	-	-	-	-	1.0	-
1-Octadecene	1791	1789	1.1	-	-	-	-	-
6,10,14-Trimethyl-2-pentadecanone	1838	1838	12.8	-	-	-	-	-
Farnesyl acetone [d]	1863	1860	2.8	-	-	-	-	-
Phytol	2112	2127	10.8	-	-	-	-	-
Total			90.1	98.2	96.7	98.3	92.0	93.1
Monoterpene hydrocarbons			2.1	14.9	53.7	40.6	34.8	8.6
Oxygenated monoterpenes			35.2	69.7	-	17.9	36.5	84.5
Sesquiterpene hydrocarbons			7.9	7.5	34.5	24.3	5.7	-
Oxygenated sesquiterpenes			17.9	-	-	15.5	15.0	-
Diterpenes			10.8	-	-	-	-	-
Aromatic compounds			6.1	6.1	8.5	-	-	-
Aliphatic compounds			9.0	-	-	-	-	-

[a] Elution order on HP-5 column; [b] Retention indices on HP-5 column; [c] Literature retention indices; [d] Correct isomer not identified; - Not identified.

Blighia sapida

The yield obtained from the hydrodistillion of *B. sapida* was 0.12% (v/w) calculated on a dry weight basis. The classes of compounds identified in *B. sapida* were oxygenated monoterpene (35.2%), oxygentaed sesquiterpenes (17.9%), diterpenes (10.8%), aliphatic compounds (9.0%) and sesquiterpene hydrocarbons (7.9%). The minor ones include monoterpene hydrocarbons (2.1%), aromatic compounds (6.1%) and fatty acids (1.1%). The main constituents of the oil (Table 1) were 6,10,14-trimethyl-2-pentadecanone or hexahydrofarnesyl acetone (12.8%), geranyl acetone (12.0%), phytol (10.8%) and α-ionone (6.1%). The minor compounds comprised of *n*-nonanal (3.6%), β-bisabolene (3.3%), farnesyl acetone (2.8%), pinocarvone (2.5%), *trans*-α-bergamotene (2.4%), naphthalene (2.4%), phenylacetaldehyde (2.2%), (Z, E)-α-farnesene (2.2%), β-ionone (2.2%) and 4-ethylguaiacol (2.1%). As mentioned earlier, no literature information is readily available on the volatile contents of this species and any other member of the genus *Blighia*.

Thevetia peruviana

The yield obtained from the hydrodistillion of *B. sapida* was 0.15% (v/w) calculated on a dry weight basis. The main classes of compounds identified in the flower essential oil of *T. peruviana* were the monoterpene hydrocarbons (14.9%) and the oxygenated monoterpenes (69.8%). The oil was devoid of any oxygenated sesquiterpene compounds (Table 1). The major constituent of the essential oil was (*E*)-β-ionone (44.5%) with significant amounts of terpinen-4-ol (8.9%), terpinolene (8.3%) and 1,8-cineole (7.1%). A sizeable quantities of 2,4-di-tert-butylphenol (6.1%) and dihydro-β-ionone (6.1%) were also identified in the oil. However, monoterpene hydrocarbons (53.8%) and sesquiterpene hydrocarbons (34.5%) represent the abundant classes of compounds present in the leaf oil. The oxygenated terpene compounds were conspicuously absent in the oil. In addition, limonene (30.0%), δ-cadinene (13.3%), α-copaene (10.1%) and terpinolene (10.0%) were the compounds occurring in higher quantity. It could be seen that the Nigerian grown *T. peruviana* oil samples does not contain fatty acids when compared with samples analyzed from Brazil [49]. However, both the Brazilian and Nigerian grown samples of *T. peruviana* contained sizeable amount of (*E*)-β-ionone and 1,8-cineole. Literature information is scanty on the chemical constituents of the leaf essential oil of *T. peruviana* and as such the present study may be the first of its kind.

Aframomum longiscapum

The percentage yields of essential oils from *A. longispacum* are 0.93% 1.13% and 0.76% (v/w) seed pod and leaf, calculated on a dry weight basis. The compounds identified in oils could be seen in Table 1. The seed comprised of monoterpene hydrocarbons (34.8%), sesquiterpene hydrocarbonse (36.5%) and oxygenated sesquiterpenes (15.0%) while oxygenated monoterpenes (84.5%) dominated in the pod. The main constituents of the seed oil **comprised mainly of** β-pinene (21.7%), *p*-cymene (13.1%), caryophyllene oxide (12.0%) and myrtenyl acetate (10.8%). The pod was rich in **1,8-cineole (54.15) and** α-terpineol (15.6%). No sesquiterpene compounds could be identified in the pod. However, **β-pinene (27.9%)**, β-caryophllene (18.8%) and caryophyllene oxide (12.2%) represents the main compounds in the leaf. There are significant **amounts of** α-pinene (9.7%), linalool (7.9%) and myrtenol (5.4%). The quantitaive amounts of β-pinene makes the composition of the oil similar to sample analysed from Cameroon sample [58], although *p*-cymene, caryophyllene oxide and mytrenal acetate were not detected in the latter.

Biological Activities of Essential Oils

Antimicrobial Assay

The results are summarized in Table 2 and Table 3 respectively for the zones of inhibition (ZI) and minimum inhibitory concentrations (MIC) values. The *B. sapida* oil displayed good antibacterial activity against the tested microorganisms. It exhibited strong activity against *B. subtilis* (ZI, 19.3 ± 2.9 mm; MIC, 0.6 mg/mL), *S. aurues* (ZI, 24.0 ± 1.0 mm; MIC 0.3 mg/mL) and *E. coli* (ZI, 21.3 ± 2.3 mm; MIC, 0.3 mg/mL), *Kliebsiella* spp. (ZI, 19.0 ± 3.6 mm; MIC, 1.3 mg/mL) and *Pseudomonas* spp. (ZI, 18.3 ± 2.9 mm; MIC 1.2 mg/mL). However, moderate activity was observed towards *Proteus* spp. (ZI, 18.0 ± 2.0 mm; MIC, 2.5 mg/mL) and *Salmonella* spp. (ZI, 16.0 ± 1.7 mm; MIC, 2.5 mg/mL) while slight resistant was displayed against *C. youagae* (ZI, 11.7 ± 2.9 mm; MIC, 5 mg/mL). The present result is in agreement with previous findings on the antimicrobial activity of different extracts of *B. sapida* [9, 10].

Table 2. Antibacterial activity of essential oils of *A. longiscapum* and *B. sapida*

Microorganisms	*A. longiscapum* [a,b] Leaf	Seed	Pod	*B. sapida* [a,b] Leaf	Gentamycin	Tetracycline
B. subtilis	16.3 ± 1.5	19.7 ± 1.5	14.7 ± 2.0	19.3 ± 2.9	20.7 ± 1.2	13.3 ± 1.2
S. aureus	18.7 ± 1.5	25.0 ± 2.0	17.0 ± 1.5	24.0 ± 1.0	21.3 ± 2.1	20.7 ± 1.5
C. youagae	11.7 ± 0.6	13.7 ± 1.5	10.7 ± 0.6	11.7 ± 2.9	8.3 ± 0.0	10.3 ± 2.1
E. coli	17.7 ± 1.5	20.3 ± 2.1	17.3 ± 1.5	21.3 ± 2.3	24.7 ± 1.2	21.7 ± 1.5
Kliebsiella spp.	14.7 ± 2.1	17.0 ± 3.0	12.3 ± 2.0	19.0 ± 3.6	10.7 ± 1.5	13.7 ± 2.1
Proteus spp.	14.3 ± 0.6	16.7 ± 0.6	12.0 ± 0.0	18.0 ± 2.0	12.3 ± 2.5	9.7 ± 1.5
Pseudomonas spp.	15.3 ± 2.3	17.3 ± 2.5	14.3 ± 1.5	18.3 ± 2.9	15.7 ± 2.1	11.0 ± 1.0
Salmonella spp.	14.0 ± 0.0	18.7 ± 0.6	13.7 ± 1.5	16.0 ± 1.7	22.0 ± 2.0	13.7 ± 0.6

Table 3. Minimum inhibitory concentrations of *A. longiscapum* and *B. sapida* essential oils [a]

Microorganisms	*A. longiscapum* Leaf	Seed	Pod	*B. sapida*	Gentamycin	Tetracycline
B. subtilis	0.6	0.3	0.6	0.6	0.3	5
S. aureus	0.6	0.1	0.3	0.3	1.2	1.2
C. youagae	5	5	5	5	> 10	5
E. coli	1.2	0.3	0.3	0.3	1.2	1.2
Kliebsiella spp.	2.5	0.6	1.2	1.3	2.5	2.5
Proteus spp.	2.5	1.2	2.5	2.5	2.5	10
Pseudomonas spp.	1.2	1.2	1.2	1.2	1.2	5
Salmonella spp.	2.5	2.5	2.5	2.5	0.6	5

[a] MIC - minimum inhibitory concentration (mg/mL).

The antimicrobial assay indicated that the leaf (ZI, 16.3 ± 1.5 mm and 18.7 ± 1.5 mm; MIC, 0.6 mg/mL), seed (ZI, 19.7 ± 1.5 mm and 25.0 ± 2.0 mm; MIC, 0.3 mg/mL and 0.1 mg/mL) and pod (ZI, 14.7 ± 2.0 mm and 17.0 ± 1.5 mm; MIC, 0.6 mg/mL) of *A. longisparcum* oil displayed strong antibacterial activity against *B. subtilis* and *S. aurues* respectively. The seed and pod oils also showed strong inhibition against *E. coli* (ZI, 17.7 ± 1.5 mm, 20.3 ± 2.1 mm and 17.3 ± 1.5 mm; MIC, 0.3 mg/mL for seed and pod; 1.2 mg/mL for leaf) while only the seed (ZI, 17.0 ± 3.0 mm; MIC, 0.6 mg/mL) was the most active towards *Kliebsiella spp*. In addition, the oil samples inhibited the growth of *Pseudomonas spp* (ZI, 15.3 ± 2.5 mm, 17.3 ± 2.5 mm and 14.3 ± 1.5 mm for leaf, seed and pod with MIC of 1.2 mg/mL) while the seed oil was most active against *Proteus spp*. (ZI, 16.7 ± 0.6 mm, MIC, 1.2 mg/mL). However, all samples displayed moderate activity towards *Salmonella spp*. (ZI, 14.0 ± 0.0 mm, 18.7 ± 0.6 mm and 13.7 ± 1.5 mm, respectively, with MIC of 2.5 mg/mL). Moreover, the leaf and pod oil samples exhibited moderate action to the growth of *Proteus spp*. (ZI, 14.3 ± 0.6 mm and 12.0 ± 0.0 mm for leaf and pod respectively, with MIC of 2.5 mg/mL).

The microbial test divulges the essential oils of *B. sapida* and *A. longiscapum* to have a broad spectrum of antibacterial activity against the tested organisms. Both *B. sapida* and *A. longiscapum* were sensitive to both gram-positive bacteria and the gram-negative bacteria. This observation is extremely remarkable because essential oils are known to be more active against gram-positive than gram-negative bacteria [69, 70].

Table 4. Insecticidal and larvicidal activities of *A. longiscapum, B. sapida* and *T. peruviana* essential oils

Sample	Part	Insecticidal [a]		Larvicidal [a]	
		% mortality	LC$_{50}$	% mortality	LC$_{50}$
A. longiscapum	Leaf	86.7	204.13 (116.37-541.54)	-	-
	Seed	93.3	117.50 (74.42-252.35)	-	-
	Pod	76.7	108.15 (55.23-321.46)	-	-
B. sapida	Leaf	100	6.28 (3.58-10.57)	100	11.61 (5.17-26.24)
T. peruviana	Flower	100	1.52 (0.00-5.87)	-	-
	Leaf	100	4.35 (1.18-13.91)	-	-
Allehtrin		100	7.45 (2.01-14.65)	-	-
Permethrin		100	11.10 (6.03-23.19)	-	-
KMnO$_4$		-	-	95.8 ± 0.3	41.37 (23.12-57.24)

[a] Lethal concentration (mg/L air) at 72 H; - Not tested.

Insecticidal Activity of Studied Essential Oils

Table 4 summaries the insecticidal and larvidical activities of A. longiscapum, B. sapida and T. peruviana against the adult of Sitophilus zeamais and fourth-instar larvae of Anopheles gambiae, respectively. In the insecticidal activity, the results from the fumigant toxicity assay showed no mortality in the negative control experiments (1% DMSO solution and pure water) after 96 H. However, at a dose of 10 µL/disc, the inhibitory activity of the essential oils of A. longiscapum, B. sapida and T. peruviana and the controls (Allethrin and Permethrin) against S. zeamais between 72 and 96 h, were found to be significant and concentration dependent. The percentage mortality of B. sapida leaf oil and T. peruviana (flower and leaf oils) at 72 h were 100%. The lethal concentrations (LC_{50}) were 6.28 mg/L air (B. sapida leaf), 1.52 mg/L (T. peruviana flower) and 4.35 mg/L air (T. peruviana leaf). The oils of A. longiscapum had percentage mortalities of 76.6 to 93.3%. The LC_{50} values were obtained as 108.15 mg/L air (pod), 204.13 mg/L air (leaf) and 117.50 mg/L air (seed). It is obvious that the oils of B. sapida and T. peruviana has strong toxic activity against S. zeamais and may be use as fumigant in protecting stored products.

Previously, extracts of B. sapida have exhibited pesticidal activity against the insect pests such as Tribolium castaneum, Callosobruchus maculatus and S. zeamais [5]. Moreover, T. peruviana plant was also reported to exhibited toxicity and fumigant activity to some insect pests [39-42].

Larvicidal Activity of B. sapida

The result of larvicidal activity of *B. sapida* leaf oil against the fourth-instar larvae of *A. gambiae* was also shown in Table 4. The larvicidal activity of the oil shows a positive correlation between the different concentrations and is directly proportional to the concentration. The investigation confirmed the toxicicity of the oil on the larvae of *A. gambiae* with lethal concentration (LC_{50}) of 11.61 mg/L air, as against 41.37 mg/L air for $KMnO_4$ used as control. Our result is in agreement with the previous findings on the extracts of the plant [3-5] and may be useful to control the spread of malaria.

Phytotoxicity Activity of T. peruviana Essential Oils

The result of phytotoxicity activity *T. peruviana* essential oils on seeds germination and seedling growth of *Zea mays* are summarized in Table 5. The results revealed the oils to have percentage germination ranging from 76.6% to 100% against the seed of *Zea mays* at 25 to 500 mg/mL in dose dependent manner. At 25 mg/mL, the flower oil has percentage germination of 83.3%, while the leaf oil had 76.7% germination inhibition. But, the flower oil at 125 mg/mL and higher, shows similar and greater percentage germination for pure water and leaf oil. Although, no significant seedling growth activities were observed, but, the radicule length of both oils displayed higher inhibition effect than plumule length and with the flower oil being more statistically noteworthy. From these results, *T. peruviana* essential oils showed phytotoxicity on the seeds of *Zea mays*, since seed emergence of radicule length (>1 mm) is regarded as germination [71-73].

The biological activity of essential oils is related in most cases to its main compounds and the synergistic effects of the minor compounds should also be taken into consideration [74]. Referring to literature, linalool, phytol and β-caryophyllene have shown antimicrobial activity. On the other hand, plant extracts and volatile oils containing sizeable proportions of phytol, α-terpineol and some other terpenes have displayed insecticidal potentials [75]. Some compound such as 1,8-cineole, limonene and (*E*)-β-ionone were known to have shown antimicrobial, insecticidal and phytotoxic activities [1, 65]. It is evident that one or more compounds present in the extracts may have been responsible for the observed insecticidal.

Table 5. **Phytotoxic effects of *T. peruviana* essential oils on seeds germination and seedling growth of *Zea mays*[a]**

Samples	Concentration	% Germination	Seedling growth (mm)	
			Radicule	Plumule
Controls	Pure water	93.3 ± 1.2	4.2 ± 2.3	2.3 ± 0.9
	1% DMSO	86.7 ± 0.6	3.7 ± 1.6	2.5 ± 1.1
Flower	25	83.3 ± 0.6	1.8 ± 0.1	0.3 ± 0.0
	50	90.0 ± 0.0	1.9 ± 0.2	0.4 ± 0.1
	125	93.3 ± 0.6	1.9 ± 0.1	0.4 ± 0.0
	250	100.0 ± 0.0	2.1 ± 0.3	0.4 ± 0.1
	500	100.0 ± 0.6	2.6 ± 0.2	0.6 ± 0.0
Leaf	25	76.7 ± 1.5	1.0 ± 0.1	0.1 ± 0.1
	50	80.0 ± 1.0	1.4 ± 0.1	0.2 ± 0.0
	125	86.7 ± 1.2	1.4 ± 0.0	0.2 ± 0.0
	250	93.3 ± 0.6	1.5 ± 0.1	0.4 ± 0.1
	500	93.3 ± 0.6	1.6 ± 0.1	0.6 ± 0.0

[a] Values are given as mean ± SD (3 replicates).

REFERENCES

[1] Lawal, O.A., Ogunwande, I.A., Bullem, C.A., Taiwo, O.T. and Opoku, A.R. (2014) Essential oil composition and *in vitro* biological activities of three *Szyzgium* species from Nigeria In: New Developments in Terpene Research (Jinnan, H. ed.), Nova Science Publisher Inc, New York, pp. 93-112, ISBN-978-1-62948-761-8.
[2] Anto, F., Aryeetey, M.E., Anyorigiya, T., Asoala, V. and Kpikpi. J. (2005). The relative susceptibilities of juvenile and adult *Bulinus globosus* and *Bulinus truncatus* to the molluscicidal activities in the fruit of Ghanaian *Blighia sapida, Blighia unijugata* and *Balanites aegyptiaca. Annal Tropical Medicine and Parasitology*, 99(2), 211-217.
[3] Adebayo, J.O and Krettli, AU. (2011). Potential antimalarials from Nigerian plants: A review. *Journal of Ethnopharmacology*, 133(2), 289-302.
[4] Ubulom, P.M.E., Imandeh, G.N., Ettebong, E.O. and Udobi, C.E. (2012). Potential larvicidal properties of *Blighia sapida*leaf extracts against larvae of *Anopheles gambiae, Culex quinquefasciatus* and *Anopheles aegypti. British Journal of Pharmaceutical Research*, 2(4): 259-268.
[5] Khan, A., Gumbs, F.A. and Persad, A. (2002). Pesticidal bioactivity of ackee (*Blighia sapida* Koenig) against three *stored-product insect pests. Tropical Agriculture*, 79(4), 217-223.
[6] Kazeem, M.I., Raimi, O.G., Balogun, R.M. and Ogundajo, A.L. (2013). Comparative study on the α-amylase andα-glucosidase inhibitory potential of different extracts of *Blighia Sapida* Koenig. *American Journal of Research Communications*, 1(7): 178-192.
[7] Gordon, A. and Jackson, J.C. (2013). The microbiological profile of Jamaican Ackees (*Blighia sapida*). *Nutrition and Food Science*, 43(2), 142-149.
[8] Hamzah, R.U., Egwim, E.C., Kabiru, A.Y. and Muazu, M.B. (2013). Phytochemical and in vitro antioxidant properties of the methanolic extract of fruits of *Blighia sapida, Vitellaria paradoxa* and *Vitex doniana. Oxidation, Antioxidant and Medical Science*, 2(3), 217-223.
[9] John-Dewole, O.O. and Popoola, O.O. (2013). Chemical, phytochemical and antimicrobial screening of extracts of *Blighia sapida* for agricultural and medicinal relevances. *Nature Science*, 11(10), 12-14.
[10] Ubulom, P.M.E., Udobi, C.E., Ekaete, A. and Udeme, E. (2013). Antimicrobial activities of leaf and stem bark extracts of *Blighia sapida. Journal of Plant Studies*, 2(2), 47-52.

[11] Oloyede, O.B., Ajiboye, T.O & Komolafe, Y.O. (2013). N-nitrosodiethylamine induced redox imbalance in rat liver: Protective role of polyphenolic extract of *Blighia sapida*. *Free Radical Antioxidant*, 3(1), 25-29.

[12] Howélé, O., Niamké, B., Dally, T. and Kati-Coulibaly, S. (2010). Nutritional composition studies of sun dried *Blighia sapida* (K. Koenig) aril from Côte d'Ivoire. *Journal of Applied Bioscience*, 32(10), 1989-1994.

[13] Emanuel, M.A., Gutierrez-Orozco, F., Yahia, E.M. and Benkeblia, N. (2013). Assessment and profiling of the fatty acids in two ackee fruit (*Blighia sapida* Köenig) varieties during different ripening stages. *Journal of Science of Food and Agriculture*, 93(4), 722-726.

[14] Omosuli, S.V. (2014). Physicochemical properties and fatty acid composition of oil extracted from ackee apple (Blighia sapida) seeds. Journal of Food Dairy Technology, 2(1), 5-8.

[15] Adeyeye, E.I. (2011). Comparability of the amino acid composition of aril and seed of Blighia sapida fruit. *African Journal of Food Agriculture and Nutritional Development, 11(3), 4810-4827.*

[16] Hassall, C.H., Reyle, K. and Feng, P. (1955). Hypoglycin A,B: Biologically active polypeptides from *Blighia sapida*. *Biochemical Journal, 60(2), 334-339.*

[17] Ainsley, P. (2007). Phytochemical analysis of ackee (*Blighia sapida*) pods. Ph.D. Thesis, City University of New York, pp. 137.

[18] Garg, H.S. and Mitra, C.R. (1967). *Blighia sapida* I. Constituents of the fresh fruit. *Planta Medica*, 15(1), 74-80.

[19] Stuart, K.L., Roberts, E.V. and Whittle, Y.G. (1976). A general method for vomifoliol detection. *Phytochemistry*, 15(2), 332-333.

[20] Mazzola, E.P., Parkinson, A., Kennelly, E.J., Coxon, B., Einbond, L.S. et al., (2011). Utility of coupled HSQC experiments in the intact structural elucidation of three complex saponins from *Blighia sapida*. *Carbohydrate Research*, 346(6), 759-768.

[21] Natalini, B., Capodiferro, V., De Luca, C. and Espinal, R. (2000). Isolation of pure (2S,1'S,2'S)-2-(2'-carbocyclopropy)glycine from *Blighia sapida* (Ackee). *Journal of Chromatography*, 873(2), 283-286.

[22] Ogunwande, I.A.& Oladosu, I.A. (2011). Anticandidal activity of phenanthrenol from *Blighia sapida* Koenig (Sapindaceae). *Der Pharmzie Lettre*, 3(1), 221-227.

[23] Frohne, D. and Pfander, H.J. (1983). A colour Atlas of poisonous plants, Germany, Wolfe Publishing Ltd., pg. 47.

[24] Panigrahi, A. and Raut, S.K. (1994) *Thevetia peruviana* (Family: Apocynaceae) in the control of slug and snail pests. *Memórias do Instituto Oswaldo Cruz*, 89(2), 247-250.
[25] Singh, A. and Singh, S. (2005). Molluscicidal evaluation of three common plants from India. *Fitoterpia*, 76(7-8), 747-751.
[26] Okoro, O., Felix, N.M., Ojimelukwe, P.C. and Ibeh, C.M. (1994). Rodenticide potential of *Thevetia peruviana*. *Journal of Herbs Spices and Medicinal Plants*, 2(3), 3-10.
[27] Sagnik, H., Indrajit, K., Mainak, C., Dilshad, A. and Pallab, H. (2015). Antitumor potential of *Thevetia peruviana* on Ehrlich's Ascites carcinoma-bearing mice. *Journal of Environment. Pathology Toxicology and Oncology*, 34(2), 105-113.
[28] Ambang, Z., Door, J.P.N., Essono, G.G., Bekolo, N., Chewachong, G.M. and Asseng C.C. (2010). Effect of *Thevetia peruviana* seeds extract on in vitro growth of four strains of *Phytophthora megakarya*. *Plant Omics*, 3(3), 70-76.
[29] Reddy, B. (2009). *Antimicrobial activity of Thevetia Peruviana (Pers.) K. Schum. and Nerium indicum Linn. The Internet Journal of Pharmacol*ogy, 8(2), 2-5.
[30] *Anupma, D., Hemlata, S., Sharma, R.A & Archana, S. (2015).* Estimation of antioxidant and antibacterial activity of crude extracts of *Thevetia peruviana* (Pers.) K. Schum. *The International Journal of Pharmacy and Pharmaceutical Sciences*, 7(5), 55-59.
[31] Radheykant, S., Priyanka, S. and Singh, V.K. (2012). Antimicrobial properties of *Thevetia peruviana*. *Rasayan Journal Chemistry*, 5(4), 403-505.
[32] Gupta, R., Kachhawa, J.B.S., Gupta, R.S., Sharma, A.K., Sharma, M.C. and Dobhal, M.P. (2011). Phytochemical evaluation and antispermatogenic activity of *Thevetia peruviana* methanol extract in male albino rats. *Human Fertility*, 14(1), 53-59.
[33] Garima, Z. and Amla, B. (2012). *In vitro* and *in vivo* determination of phenolic contents and antioxidant activity of desert plants of Aapocynaceae family. *Asian Journal of Pharmacy and Clinical Research*, 5(Suppl. 1), 76-83.
[34] Pavithra, G.S., Anusha, M and Rajyalakshmi, M. (2012). Effect of *Thevetia peruviana* extracts on I and in-vivo cultures of *Parthenium hysterophorus* L. *Journal of Crop Science*, 3(3), 83-86.
[35] Komal, A. (2013). Allelopathic effect of *Thevetia peruviana* on growth of *Triticum aestivum*. *Indian Journal of Plant Science*, 2(4), 10-13.

[36] Tewtrakul, S., Miyashiro, H., Nakamura, N., Hattori, M., Kawahata, T., Otake, T., et al., (2003). HIV-1 integrase inhibitory substances from *Thevetia peruviana*. *Phytother Research*, 17(3), 232-239.
[37] Sunil, K.S., Shailendra, K.S. and Ajay, S. (2013). Toxicological and biochemical alterations of apigenin extracted from seed of *Thevetia peruviana*, a medicinal plant. *Journal of Biology and Earth Science*, 3(1), B110-B119.
[38] Ama, T.T., Kwadwo, T.W., Bosu, P.P. and Hawkins, E.J. (2014). The termite controlling capabilities of extracts of *Thevetia peruviana* (Pers.) K. Schum in Ghana. *Interlink Continental Journal Environmental Science and Toxicology*, 1(1), 001-008.
[39] Singh, S.K., Yadav, R.P. and Singh, A. (2010). Piscicidal activity of leaf and bark extract of *Thevetia peruviana* plant and their biochemical stress response on fish metabolism. *European Review Medical and Pharmacology Science*, 14(11), 915-923.
[40] Umollah, J.U. and Islam, D.W. (2007). Toxicity of *Thevetia peruviana* (Pers) Schum. extract to adults of *Callosobruchus maculatus* F. (Coleoptera: Bruchidae). *Journal of Agriculture and Rural Development*, 5(1&2), 105-109.
[41] Sathish, V., Umavathi, S., Thangam, Y. and Mathivanan, R. (2015). Analysis of phytochemical components and larvicidalactivity of *Thevetia peruviana* (Pers) Merr, against the Chickungunya vector *Aedes aegypti* (L). *Internal Journal of Current Microbiology and Applied Science*, 4(1), 33-39.
[42] Suresh, Y., Singh, S.P. and Mittal, P.K. (2013). Toxicity of *Thevetia peruviana* (yellow oleander) against larvae of *Anopheles stephensi* and *Aedes aegypti* vectors of malaria and dengue. *Journal of Entomology and Zoological Studies*, 1(6), 85-87.
[43] Cruz, W.T., Silva, M.Z.R., Moreno, F.B.M.B., Monteito-Moreira, A.C.O., Vieira-Neto, A.E., Souza, P.F.N., et al., (2015). Proteomic profile from the latex of *Thevetia peruviana*. 23rd Congress of the International Union for Biochemistry and Molecular Biology 44th Annual Meeting of the Brazilian Society for Biochemistry and Molecular Biology, Foz do Iguaçu, PR, Brazil, pg. 122.
[44] Alabi, K.A., Lajide, L. and Owolabi, B.J. (2013). Analysis of fatty acid composition of *Thevetia peruviana* and *Hura crepitans* seed oils using GC-FID. *Fountain Journal of Natural and Applied Science*, 2(2), 32-37.

[45] Kohls, S., Scholz-Böttcher, B.M., Jörg, T., Patrick, Z. and Jürgen, R. (2012). Cardiac glycosides from yellow oleander (*Thevetia peruviana*) seeds. *Phytochemistry*, 75, 114-127.

[46] Dan-Mei, T., Huo-Yun, C., Miao-Miao, J., Wei-Zai, S., Jin-Shan, T. and Xin-Sheng. Y. (2016). Cardiac glycosides from the seeds of *Thevetia peruviana*. *Journal of Natural Product*, 79(1), 38-50.

[47] Thilagavathi, R., Kavitha, H.P. and Venkatraman, B.R. (2010). Isolation, characterization and anti-inflammatory property of *Thevetia Peruviana*. *E-Journal Chemistry*, 7(4), 1584-1590.

[48] Tewtrakul, S., Nakamura, N., Hattori, M., Fujiwara, T. and Supavita, T. (2002). Flavanone and flavonol glycosides from the leaves of *Thevetia peruviana* and their HIV-1 reverse transcriptase and HIV-1 integrase inhibitory activity. *Chemical and Pharmaceutical Bulletin*, 50(5), 630-635.

[49] Pragati, K., Prabodh, S., Padmini, S. and Shashi, A. (2015). Antinociceptive and gastro protective effects of inhaled and orally administered *Thevetia peruviana* Pers. K. Schum essential oil. *International Journal of Pharmaceutical Science Research*, 6(10), 4496-4502.

[50] Ballabeni, V., Tognolini, M., Chiavarini, M., Impicciatore, M., Bruni, R., Bianchi, A, et al. (2004). Novel antiplatelet and antithrombotic activities of essential oil from *Thevetia peruviana* Pers. K. Shum. *Phytomedicine*, 67(12), 234-239.

[51] Maia, J.G.S., das Graças, M.B.Z., Andrade, E.H.A. and Carreira, L.M.M. (2000). Volatiles from flowers of *Thevetia peruviana* (Pers.) K. Schum. and *Allamanda cathartics* Linn. (Apocynaceae). *Journal of Essential Oil Research*, 12(3), 322-324.

[52] Hepper, F.N. (2000). Flora of West Tropical Africa: A descriptive account of Butomaceae-Orchidaceae in West Tropical Africa, Royal Botanical Garden, Kew: Vol 3, Part 1, pp. 276.

[53] Muhammad, A., Olagunju, A. and Ndidi, U.S. (2012). Nutrients and anti-nutrients analyses of *Aframomum longiscapum* seeds. *International Journal of Food Nutrition and Safety*, 1(3), 120-126.

[54] Nwozo, S.O., Okameme, P.E. and Oyinloye, B.E. (2012). Potential of *Piper guineense* and *Aframomum longiscapum* to reduce radiation induced hepatic damage in male Wistar rats. *Radiats Biology and Radioecology*, 52(4), 363-369.

[55] Muhammad, A., Habila, N., Aimola, I.A., Olagunju, A., Egbeogu, V.O., Edet, I, et al. (2012). In vitro antioxidant potentials of extracts of

Aframomum longiscapum seed. *International Journal of Traditional and Natural Medicine*, 1(1), 8-19.

[56] Owumi, S.E., Oyeronke, A.O. and Mohammed, A. (2012). Co-administration of sodium arsenite and ethanol: Protection by aqueous extract of *Aframomum longiscapum* seeds. *Pharmacognosy Research*, 4(3), 154-160.

[57] Soraya, A., Victor, N., Neil, G.H., Simon, O. and Michel, I. (1999). *Aframonum* seeds for improving penile activity. US Patient No. 5879682 A.

[58] Diomandé, G.D., Koffi, A.M., Tonzibo, Z.F., Bedi, G. and Figueredo, G. (2012). GC and GC/MS analysis of essential oil of five *Aframomum* Species from Côte D'ivoire. *Middle-East Journal of Scientific Research*, 11(6), 808-813.

[59] British Pharmacopoeia specifications. (1980). HM Stationary Office, Vol. II.

[60] *National Institute of Standards and Technology. (2011).* Chemistry web book. Data from NIST Standard Reference Database 69 (http://www.nist.gov/).

[61] Vijoen, A.M., van Vuuren, S.F., Gwebu, T., Demirci, B. and Baser, K.H.C. (2006). The geographical variation and antimicrobial activity of African wormwood *(Artemisia afra* Jacq.) essential oil. *Journal of Essential Oil Research*, 18(1), 19-25.

[62] Eloff, J.N. (1998). A sensitive and quick microplate method to determine the minimal inhibitory concentration of plant extracts for bacteria. *Planta Medica*, 64(8), 711-713.

[63] Hashemi, S.M. and Safavi, S.A. (2012). Chemical constituents and toxicity of essential oils of Oriental arborvitae, *Platycladus orientalis* (L.) Franco, against three stored-product beetles. *Chilean Journal of Agricultural Research*, 72(2), 188-194.

[64] Cetin, H. and Yanikoglu, A. (2006). A study of the larvicidal activity of *Origanum* (Labiatae) species from southwest Turkey. *Journal of Vector Ecology*, 31(1),118-122.

[65] Cheng, S., Liu, J., Tsai, K., Chen, W. and Chang, S. (2004). Chemical composition and mosquito larvicidal activity of essential oils from leaves of different *Cinnamomum osmophloeum* Provenances. *Journal of Agriculture and Food Chemistry*, 52(14), 4395-4400.

[66] Prajapati, V., Tripathi, A.K., Aggrawal, K.K. and Khanuja, S.P.S. (2005). Insecticidal, repellent and oviposition-deterrent activity of selected

essential oils against *Anopheles stephensi, Aedes aegypti* and *Culex quinquefasciatus*. *Bioresources Technology*, 96(16), 1749-1757.
[67] Faravani, M.H.B., Baki, H.B. and Khalijah, A. (2008). Assessment of allelopathic potential of *Melastoma malabathricum* L. on Radish (*Raphanus sative* L) and Barnyard grass (*Echinochloacrus-galli*). *Notany Botanay and Horticutural Agrobotany Cluj*, 36(2), 54-60.
[68] Abbott, W.S. (1925). A method for computing the effectiveness of an insecticide. *Journal of Economy and Entomology*, 18(3), 265-267.
[69] Keawsa-Ard, S., Liawruangrath, B., Liawruangrath, S., Teerawutgulrag, A. and Pyne, S.G. (2012). Chemical constituents and antioxidant and biological activities of the essential oil from leaves of *Solanum spirale*. *Natural Product Communications*, 7(7), 955-958.
[70] Saha, S., Dhar, N.T., Sengupta, C. and Ghosh, P. (2013). Biological activities of essential oils and methanol extracts of five *Ocimum* species against pathogenic bacteria. *Czech Journal of Food Science*, 31(2), 194-202.
[71] Moore, M.T., Huggett, D.B., Huddleston, G.M., Rodgers, J.H. and Cooper, C.M. (1999). Herbicide effects on *Typha latifolia* (Linneaus) germination and root and shoot development. *Chemosphere*, 38(15), 3637-3647.
[72] Munzuroglu, O. and Geckil, H. (2002). Effects of metals on seed germination, root elongation, and coleoptile and hypocotyl growth in *Triticum aestivum* and *Cucumis sativus*. *Archives of Environment Contaminant and Toxicology*, 43(2), 203-213.
[73] Murata, M.R., Hammes, P.S. and Zharare, G.E. (2003). Effect of solution pH and calcium concentration on germination and early growth of groundnut. *Journal of Plant Nutrition*, 26(6), 1247-1262.
[74] Lawal, O.A., Ogunwande, I.A., Salvador, A.F., Sanni, A.A. and Opoku, A.R. (2014). *Pachira glabra* Pasq. Essential oil: Chemical constituents, antimicrobial and insecticidal activities. *Journal of Oleo Science*, 63(3), 629-635.
[75] Lawal, O.A., Opoku, R.A. and Ogunwande, I.A. (2015). Phytoconstituents and insecticidal activity of different leaf solvent extracts of *Chromolaena odorata* against *Sitophilus zeamais*. *European Journal of Medicinal Plants*, 5(3), 237-247.

Reviewed by: Dr. Akintayo L. Ogundajo, Phytomedicine and Phytopharmacology Research Laboratory, Faculty of Natural and

Agricultural Science, University of Free state, Qwaqwa Campus, South Africa. Tel:+27788796568; E-mail;ogundajotayo@yahoo.com

BIOGRAPHICAL SKETCH

Name: Oladipupo AdejumobiLawal

Affiliation: Department of Chemistry, **Faculty of Science,** Lagos State University, Ojo campus, PMB 0001, LASU Post Office, Lagos State, Nigeria.

Education: (i) Lagos State University, Ojo Lagos State, Nigeria – B.Sc (Hons) Chemistry
(ii) University of Lagos, Akoka, Yaba, Nigeria – M.Sc Chemistry (iii) University of Zululand, KwaDlangezwa, South Africa – Ph.D Chemistry

Business Address: Natural Products Research Unit, Department of Chemistry, **Faculty of Science,** Lagos State University, Ojo campus, PMB 0001, LASU Post Office, Lagos State, Nigeria. jumobi.lawal@lasu.edu.ng

Research and Professional Experience:
Postdoctoral research experience at the Department of Biochemistry and Microbiology, University of Zululand, KwaDlangezwa, South Africa

Reviewer to:
American Association for Science and Technology, Drug Research, European Journal of Medicinal Plants, Journal of Environmental Health Science and Engineering, Journal of Essential Oil-Bearing Plants, Journal of Herbs, Spices & Medicinal Plants, Journal of Medicinal Plant Research, Natural Product Communications, Natural Product Research, Pharmaceutical Biology, The African Journal of Plant Science

Professional Appointments:
External Examiner - Ph.D Thesis Evaluation, Department of Biochemistry and Microbiology, University of Zululand, KwaDlangezwa, South Africa.

In: Advances in Chemistry Research. Volume 35 ISBN: 978-1-53610-734-0
Editor: James C. Taylor © 2017 Nova Science Publishers, Inc.

Chapter 7

FLUORESCENCE AND PHOSPHORESCENCE SPECTRA OF XANTHONE IN THE VAPOR PHASE

Takao Itoh[*]

Graduate School of Integrated Arts and Sciences, Hiroshima University,
1-7-1 Kagamiyama, Higashi-Hiroshima City, 739-8521, Japan

ABSTRACT

Spectral measurements of molecules in the vapor phase provide intrinsic properties of molecules free from the influence of environment. Emission spectra of xanthone vapor measured at different temperatures are shown along with the excitation and absorption spectra. The emission consists of fluorescence from S_1 (n, π^*), $^1A_2$ and phosphorescence from T_1 (n, π^*), $^3A_2$ state overlapping each other. The overlapping spectrum was separated to extract only each of the fluorescence and phosphorescence spectra. The vibrational structures of both the fluorescence and phosphorescence were interpreted in terms of the C = O stretching mode and the modes combined with it. The S_1 and T_1 origins are located at 26940 and 25700 cm^{-1}, respectively. Analysis of the data

---

[*] Professor Emeritus of Hiroshima University, Japan; Present address: 1-30-29 Kamishakujii, Nerima-ku, Tokyo, 177-0044 Japan.

includes the determination of the vibrational frequencies in the fluorescence and phosphorescence spectra.

1. INTRODUCTION

Up to now a number of spectroscopic data have been accumulated for xanthone, but the origin of the emission in the condensed phases seems to be still in controversy [1 - 10]. The discussion on the excited states of xanthone has been focused mainly on the relative energy levels of the two closely located $^3(\pi, \pi^*)$ and $^3(n, \pi^*)$ states [1 - 11]. This molecule is reported to exhibit dual phosphorescence from the closely located T_2 and T_1 states [1 - 4]. Thermally activated T_2 phosphorescence in polyethylene films was also reported for xanthone [12]. However, it is likely that most of the observed dual phosphorescence can be attributed to intermolecular interactions, site effects or difference of the molecular geometry [1 - 3, 5, 11].

In order to reveal the emission properties and nature of the excited states of xanthone, it is of importance to investigate the emission in the vapor phase where the influence of environment is neglected. Due to the relatively small S_1 - T_1 energy separation [13] the relative intensity of the thermally activated delayed fluorescence of xanthone vapor is expected to be high as compared with other aromatic carbonyl molecules such as benzaldehyde. To the best of our knowledge, the emission spectrum of xanthone vapor has not been reported until quite recently.[14] Further, only a small number of examples are known for the observation of the fluorescence from the $^1(n, \pi^*)$ state of C = O containing molecules [15].

In the present chapter, emission spectra are shown for xanthone vapor measured at different temperatures from 38 °C up to 159 °C, along with the excitation and absorption spectra. Fluorescence spectrum from the $S_1(n, \pi^*)$ state of xanthone vapor was extracted by subtracting the emission spectrum measured at lower temperatures from that measured at higher temperatures.

Conversely, vapor-phase phosphorescence from the $T_1(n, \pi^*)$ state also was extracted by subtracting the emission spectrum measured at higher temperatures from that measured at lower temperatures. The vibrational structures of the fluorescence and phosphorescence were interpreted in terms of the C = O stretching mode accompanied by the modes combined with it. The locations of the S_1 and T_1 origins are observed at 26940 and 25750 cm^{-1}, respectively. Analysis of the data includes the determination of the vibrational frequencies of the bands appearing in the fluorescence spectrum. Experimental details for the sample preparation and spectral measurements are described in a paper [14].

2. EMISSION AND ABSORPTION SPECTRA IN THE CONDENSED PHASES

Absorption spectra in non-polar saturated hydrocarbons at temperatures near room temperature provide basic information on the electronic excited states of molecules. Such spectra are normally easy to measure, with the spectral pattern essentially resembling that obtained in the vapor phase. Figure 1 shows absorption spectrum of xanthone in hexane at room temperature. It is seen that there is a very weak band located at 25750 cm^{-1}. This weak band is assigned to the origin of the $S_0 \rightarrow T_1$ (n, π^*) absorption band with the ε value of about 0.5, which is a reasonable value for the $S_0 \rightarrow T_1$ (n, π^*) transition. The origin of the $S_0 \rightarrow S_1$ (n, π^*) absorption band, seen at the shoulder of the strong $S_0 \rightarrow S_2$ (π, π^*) absorption, is located at 27200 cm^{-1} with the ε value of about 20, which is also a reasonable value for the $S_0 \rightarrow S_1$ (n, π^*) transition. Thus, the S_1 - T_1 energy separation of xanthone is obtained to be about 1450 cm^{-1} in hexane. The origin of the $S_0 \rightarrow S_2$ absorption is seen at 29850 cm^{-1} with ε value of 11800.

Phosphorescence spectrum of xanthone in a (1: 1) mixture of isopentane and methylcyclohexane measured at 77 K is shown in Figure 2. This spectrum shows a prominent progression in the C=O stretching vibration along with the bands combined with it, showing a character of the phosphorescence from the ^{3}(n, π^*) state. The origin band is seen at 25800 cm^{-1}, which agrees well with that observed in the absorption spectrum in hexane (see Figure 1). In a case where the $^3(\pi, \pi^*)$ state is located at energies below the ^{3}(n, π^*) state, the phosphorescence emission should occur from the lowest $^3(\pi, \pi^*)$ state in a glass at 77 K. Such situations have been observed for aromatic carbonyl molecules such as *p*-cyanobenzaldehyde or *p*-methoxybenzaldehyde.[16, 17] Thus, the lowest excited triplet state of xanthone in an isopentane - methylcyclohexane mixed glass at 77 K is safely regarded as of the $^3(n, \pi^*)$ type.

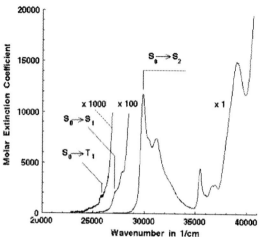

Figure 1. Absorption spectrum of xanthone in hexane at room temperature.

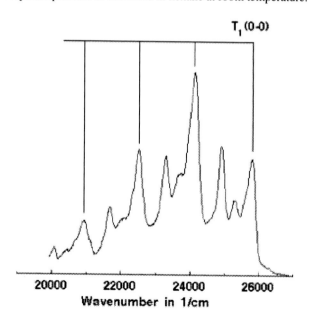

Figure 2. Phosphorescence spectrum of xanthone in an isopentane-methylcyclohexane (1:1) mixture at 77 K. The C=O stretching progression is indicated by solid lines.

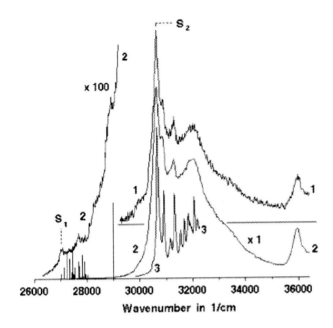

Figure 3. Absorption (1) and excitation spectra (2 and 3) of xanthone vapor: In the static vapor phase (1 and 2), and in a jet (3). The sticks below 30000 cm^{-1} indicate the locations of the peaks in the $S_0 \to S_1$ excitation spectrum in a jet. Jet data are taken from Ref. 19. The excitation spectrum in the static vapor phase was corrected.

3. EMISSION, EXCITATION AND ABSORPTION SPECTRA IN THE VAPOR PHASE

Figure 3 shows the absorption and corrected excitation spectra of xanthone vapor. It is seen that the excitation spectrum agrees well with the absorption spectrum. Further, there are also good correspondences between the excitation spectrum in a jet and that in the static vapor phase. The S_1 (n, π*) absorption origin is seen at 26940 cm^{-1}. Hirota et al. reported the excitation spectrum of xanthone vapor in a jet and assigned the band at 26939 cm^{-1} as the S_1 origin [18]. The band at 29383 cm^{-1} had been assigned initially as the S_2 origin [18], but later the band at 30718 cm^{-1} was reassigned as the S_2 origin [19]. Indeed, the S_2 origin reassigned later agrees in position reasonably with that observed in room-temperature solutions.

Figure 4 shows emission spectra of xanthone vapor measured at different temperatures. The emission whose relative intensities increase with increasing temperature corresponds to the fluorescence. On the other hand, emission bands whose relative intensities decrease with increasing temperature correspond to the phosphorescence. It is seen in Figure 4 that the emission spectral pattern changes significantly depending on temperature and that the relative intensity of the fluorescence increases with increasing temperature, showing a character of thermally activated delayed fluorescence. Thus, the logarithmic values of the fluorescence/phosphorescence intensity or quantum yield ratios (I_F/I_P or Φ_F/Φ_P) plotted against the reciprocal of the absolute temperature ($1/T$) give a linear relationship, due to the relation, $\ln\Phi_F/\Phi_P \cong \ln k_F/k_P - \Delta E/k_B \times 1/T$, where k_B, k_F and k_P are, respectively, the Boltzmann constnt, radiative rate constants of the S_1 and T_1 states. The slope of the plots provides the $S_1 - T_1$ energy difference (ΔE) of 1260 ± 50 cm^{-1}, which is very close to the spectroscopically determined separation of 1240 cm^{-1}.

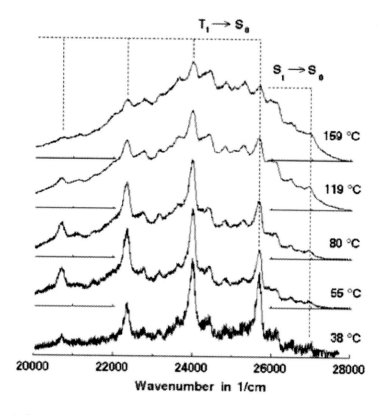

Figure 4. Corrected emission spectra of xanthone vapor at different temperatures. The C = O stretching progression in the phosphorescence is indicated by broken lines.

Most part of the emission measured at 159 °C is considered to consist of the fluorescence, while that measured at 38 °C consists mostly of the phosphorescence. The S_1 fluorescence origin is observed at 26940 cm^{-1}, which agrees well with the S_1 absorption origin. The strong peak seen at 25700 cm^{-1} in the emission spectrum measured at 38 °C is assigned to the T_1 (n, π*) origin. The location of the T_1 origin (25700 cm^{-1}) is somewhat lower in energy than that determined from the excitation spectrum in a jet (25808 cm^{-1}),[18] but is very close to the value evaluated from the phosphorescence peak in n-hexane at low temperatures (25720 or 25750 cm^{-1}) [1, 8] as well as to the location of the weak $S_0 \rightarrow T_1$(n, π*) absorption band at 25750 cm^{-1} in n-hexane at room temperature shown in Figure 1.

In order to obtain the emission spectrum that consists almost of the phosphorescence, the emission spectrum measured at higher temperature, after multiplying an appropriate number, was subtracted from that measured at lower temperature. Then, the subtracted emission spectrum, also after multiplying an appropriate constant, was subtracted from that measured at higher temperature to obtain the emission that consists mostly of the fluorescence. These procedures were repeated to extract only the fluorescence spectrum. The extracted phosphorescence spectrum consists of mostly the single mode, showing prominent progression in the C = O stretching vibration. This situation is similar to the phosphorescence spectrum observed in n-hexane at 77 K, where only the C = O stretching mode appears [8]. Figure 5 shows the extracted delayed fluorescence spectrum of xanthone vapor. The vibrational structure in the fluorescence spectrum was interpreted in terms of the C = O stretching mode and the modes combined with the C = O stretching, showing a maximum at near 24000 cm^{-1} in the emission envelop. Further, the extracted fluorescence forms a sort of mirror image relationship with the $S_0 \rightarrow S_1$ excitation spectrum. Figure 6 shows the extracted phosphorescence spectrum of xanthone vapor, which resembles the emission measured at low temperature as well as the phosphorescence of other C = O containing molecules such as benzaldehyde in the vapor phase.

The locations and assignments of the peaks observed in the fluorescence and phosphorescence spectra of xanthone vapor are summarized in Table 1 along with brief assignments. DFT-calculated wavenumbers in the 200 ~ 1700 cm^{-1} region are also given in Table 1 along with reported infrared data [20]. The optimized geometry of xanthone obtained by DFT calculation at the B3LYP/6-31++G(d,p) level is planar with the C_{2v} symmetry. The 63 fundamentals of xanthone vibration are divided into 22 of the a_1 (in-plane modes), 9 of the a_2 (out-of-plane), 11 of the b_1 (out-of-plane) and 21 of the b_2

(in-plane) species. It is seen in Table 1 that the C=O stretching frequencies of the 0-0, 0-1, 0-2 and 0-3 bands in the phosphorescence and fluorescence spectra provide exactly the same value (~ 1675 cm^{-1}), reflecting identical frequencies in the ground state.

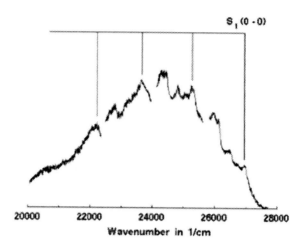

Figure 5. Extracted fluorescence spectrum of xanthone vapor. The C=O stretching progression is indicated by solid lines.

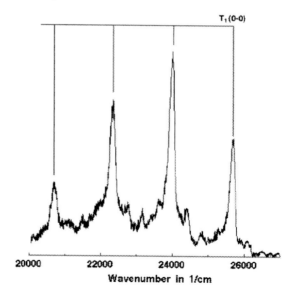

Figure 6. Extracted phosphorescence spectrum of xanthone vapor. The C = O stretching progression is indicated by solid lines.

Table 1. Frequencies (ν) in cm^{-1} and assignments of the bands observed in the fluorescence and phosphorescence spectra of xanthone vapor

ν [a]	Fluorescence	Phosphorescence	DFT[b]	IR	Assignment
26940	0				S1 ($^1A_2$) Origin
26705	235		233.2	235	Skeletal bending (a1)
26495	445		446.5	444	Skeletal deform. (b2)
26100	840		844.8	843	Skeletal deform. (a1)
25915	1025		1037.9	1030	Ring deform. (a1)
25700		0			T1 ($^3A_2$) Origin
25265	1675		1683.8	1668	C=O Stretching (a1)
25030	1910				1675 + 235
24860		840	844.8	843	Skeletal deform. (a1)
24820	2120				1675 + 445
24425	2515				1675 + 840
24240	2700				1675 + 1025
24025		1675	1683.8	1668	C=O Stretching (a1)
23600	3340				1675 x 2
23365	3575				1675 x 2 + 235
23185		2515			1675 + 840
22760	4180				1675 x 2 + 840
22575	4365				1675 x 2 + 1025
22360		3340			1675 x 2
21945	4995				1675 x 3
21520		4180			1675 x 2 + 840
20705		4995			1675 x 3

[a] Accuracy of ± 5 cm^{-1}.
[b] DFT calculation at the B3LYP/6-311++G** level.
[c] Infrared (IR) data in solution taken from Ref. 20.

SUMMARY

Fluorescence from the S_1 (n, π*), $^1A_2$ and phosphorescence from the, T_1 (n, π*), $^3A_2$ states of xanthone vapor were separated to extract only the fluorescence or phosphorescence spectrum. The vibrational structures of the extracted fluorescence and phosphorescence was interpreted in terms of the C = O stretching mode and the modes combined with that mode. The locations of the S_1, $^1A_2$ and T_1, $^3A_2$ origins are observed at 26940 and 25700 cm^{-1}, respectively.

REFERENCES

[1] H. J. Pownall, and J. R. Huber, J. Am. Chem. Soc. 93 (1971) 6429 - 6436.
[2] H. J. Pownall, R. E. Connors, and J. R. Huber, *Chem. Phys. Lett.* 22 (1973) 403 - 405.
[3] M. Vala, J. Hurst, and I. Trabjerg, Mol. Phys. 43 (1981) 1219 - 1234.
[4] R. E. Connors, R. J. Sweeney, and F. Cerio, *J. Phys. Chem.* 91 (1987) 819 - 822
[5] H. J. Pownall, and W. W. Mantulin, *Molec. Phys.* 31 (1976) 1393 - 1406.
[6] R. E. Connors, and P. S. Walsh, *Chem. Phys. Lett.* 52 (1977) 436 - 438.
[7] A. Chakrabarti, and N. Hirota, *J. Phys. Chem.* 80 (1976) 2966 - 2973.
[8] H. J. Griesser, and R. Bramley, *J. Lumin.* 24/25 (1981) 531 - 534.
[9] H. J. Griesser, and R. Bramley, *Chem. Phys. Lett.* 83 (1981) 287 -291.
[10] H. J. Griesser, and R. Bramley, *Chem. Phys. Lett.* 88 (1982) 27 -32.
[11] T. Itoh, *Chem. Rev.*, 112 (2012) 4541 - 4568.
[12] R. E. Connors, R. J. Sweeney, and F. Cerio, *J. Phys. Chem.* 91 (1987) 819 - 824.
[13] V. Rai-Constapel, M. Etinki, and C. M. Marian, *J. Phys. Chem.* A 117 (2013) 3935 - 3994.
[14] T. Itoh, *Chem. Phys. Lett.* 559 (2014) 12 - 14.
[15] M. A. El-Sayed, Acc. *Chem. Res.* 1 (1968) 8 - 16.
[16] T. Itoh, J. Lumin. 109 (2004) 221 - 225.
[17] T. Itoh, Chem. *Phys. Letters*, 591 (2014) 109 - 112.
[18] M. Baba, T. Kamei, M. Kiritani, S. Yamauchi, and N. Hirota, *Chem. Phys. Lett.* 185 (1991) 354 - 358.
[19] Y. Ohshima, T. Fujii, T. Fujita, D. Inaba, and M. Baba, *J. Phys. Chem.* A 107 (2003) 8851 - 8855.
[20] R. Zwarich, and O. S. Binbrek, Spectrochim Acta 41A (1985) 537 - 544.

In: Advances in Chemistry Research. Volume 35   ISBN: 978-1-53610-734-0
Editor: James C. Taylor   © 2017 Nova Science Publishers, Inc.

Chapter 8

MASS TRANSFER CHARACTERISTICS OF AN EXTERNAL LOOP AIRLIFT REACTOR

Makaira Govender, Christopher Perumal, Elley M. Obwaka and Amir H Mohammadi**
Discipline of Chemcal Engineering, School of Engineering,
University of KwaZulu-Natal, Howard College Campus,
Durban, South Africa

ABSTRACT

Airlift reactors facilitate contact of a liquid with a gas or solid phase while providing good agitation, mass and heat transfer. They are applied in both chemical and biochemical industries. The aim of this study was to determine and compare the overall volumetric mass transfer characteristics for an external loop airlift reactor for four configurations, taking into account hydrodynamic properties such as gas hold up, superficial gas velocity, superficial liquid velocity and pressure drops. A total of twenty runs were conducted, five runs per configuration, with varying gas flowrate for each run from 0.5l/s to 2l/s. Flow rates above these values caused spillage from the disengagement tank. For each run, manometer readings and liquid flow rates were recorded in addition to dissolved oxygen concentrations obtained using a dissolved oxygen probe and YSI Data Manager Software. Thereafter, gas hold up, superficial gas

---

* Corresponding authors: M.G. Ntunka: Ntunka@ukzn.ac.za and A.H. Mohammadi: a.h.m@irgcp.fr & amir_h_mohammadi@yahoo.com.

and liquid velocity and overall volumetric mass transfer coefficient were calculated. The results acquired showed distinct trends. Increasing the superficial gas velocity increased the gas hold up, liquid circulation velocity and overall volumetric mass transfer coefficient in a linear manner. Configuration 1, only the riser, displayed the best results with a gas hold up of 0.8701 and overall volumetric mass transfer coefficient of $0.0420s^{-1}$. Downcomer 1 in configuration 3 had the highest liquid circulation velocity of 1.41m/s as it had the smallest diameter. In general, configuration 1 showed the best results which can be attributed to the sparger being placed at the base of the riser. This allowed for direct contact of the gas with no bubbles disengaging in the disengagement tank, however, the downcomers are useful for a greater degree of mixing which was not achieved in the riser. Results are in accordance with literature trends. Outliers in data points were caused by deviations in flow rates used and a change in flow regime over a superficial gas velocity of 0.06m/s.

NOMENCLATURE

Symbol	Description	Unit
A_{Di}	Cross-Sectional Area of Downcomer	m^2
A_R	Cross-Sectional Area of Riser	m^2
d_{Di}	Diameter of Downcomer	m
d_R	Diameter of Riser	m
DO	Dissolved Oxygen Concentration	kg/m^3
DO^0	Initial Dissolved Oxygen Concentration	kg/m^3
DO^*	Concentration of Saturated Dissolved Oxygen	kg/m^3
U_G	Superficial Gas Velocity	m/s
U_{GDi}	Superficial Gas Velocity In Downcomer	m/s
U_{GR}	Superficial Gas Velocity In Riser	m/s
U_L	Superficial Liquid Velocity	m/s
U_{LDi}	Superficial Liquid Velocity In Downcomer	m/s
U_{LR}	Superficial Liquid Velocity In Riser	m/s
Q_i	Volumetric Flow Rate	l/s
$Äh_m$	Differential Heights	m
$Äz$	Difference in height Between Sampling Points	m
ε_G	Gas Holdup	Dimensionless
ρ_i	Density	kg/m^3

INTRODUCTION

Airlift reactors (ALRs) may be simply defined as units that facilitate contact with a gas, liquid or solid and are primarily used in processes that require both rapid and uniform distribution of components with good mass and heat transfer (Jin, Yin and Lant, 2006). They induce circulation by gas-lift action due to varying bulk densities (Chisit, 1987). The external loop airlift reactor, of particular interest in this study, is characterised by fluid flow through distinct conduits. ALRs are very similar to bubble column reactors, however, they are favoured due to their ability to achieve increased liquid circulation velocity and offer higher degree of turbulence, thus resulting in higher mass and heat transfer rates. In general, ALRs are advantageous over other reactors, such as stirred tank reactors, due to their ease of construction with no moving parts, simple scale up, low energy requirements and low shear rate (Zhonghuo, Tiefeng, Nian and Zhanwen, 2010).

ALRs are commonly used in chemical and biochemical processes such as production of beer, citric acid and vinegar, PRUTEEN processes, aerobic fermentation, culturing of plant and animal cells and treatment of municipal and industrial wastewater.

This study focuses on determining and comparing the overall volumetric mass transfer coefficients for a two phase air-water system of an external loop airlift reactor for four different loop configurations at a constant level of liquid. Hydrodynamic properties of importance include liquid circulation velocity, superficial gas velocity, gas holdups and pressure drops.

Previously, studies have been conducted on small scale airlift reactors of 2-4m in height (Zhonghuo, Tiefeng, Nian and Zhanwen, 2010) thus investigations on a larger ALR of 6m in height will prove useful in industry for scale up procedures as size plays an important role with regards to hydrodynamic properties and rate of mass transfer.

THEORETICAL BACKGROUND

Airlift Reactors (ALRs) are used in chemical and biochemical processes that require rapid and uniform distribution of reaction components. They can be used for multiphase systems in which good heat and mass transfer characteristics are a necessity (Zhonghuo, Tiefeng, Nian and Zhanwen, 2010).

The air or gas used in ALRS (or gas lift reactors for the latter case) facilitates agitation and transfer of heat and mass between phases in the reactor.

CLASSIFICATION

ALRs can be categorised into two main groups, internal and external loop airlift reactors, based solely on their structure. External loop airlift reactors circulate fluids via separate and distinct conduits whereas internal loop airlift reactors circulate fluid through a single vessel with baffles/draught tubes that produce separate channels for circulation (Merchuk & Gluz, 2000).

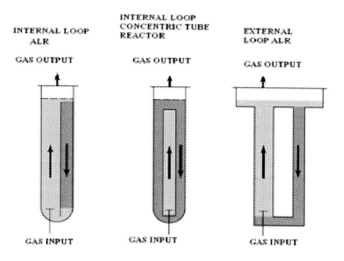

Figure 1. Various Types of Loop Airlift Reactors (Merchuk and Siegel, 1988).

ALRs (both internal and external loop) consist of four main regions which have distinct characteristics in terms of momentum, mass and heat transfer.

- Riser: A sparger is located at the bottom of this section through which gas is injected. Flow of both the gas and liquid is upward in this region.
- Downcomer: The downcomer is located parallel to the riser. It is connected at both ends to the riser thus forming a continuous loop through which fluid flow is predominately downward.

- Base: The base has negligible effect on hydrodynamic behaviour of the reactor, it simply connects the riser and downcomer at the bottom.
- Gas Separator/Disengagement Tank: This section is located at the top of the ALR where the riser and downcomer are connected. It facilitates gas disengagement and liquid recirculation. In this region, increasing the gas residence time decreases the fraction of gas recirculating through the downcomer (Merchuk & Siegel, 1988).

Sparger Classification

An instrumental design consideration of the ALR is the sparger as it plays a major role in the hydrodynamic behaviour of the reactor. It affects the liquid circulation velocity and gas hold up and mixing. Parameters that affect these aspects include location of the sparger, orifice diameter and free area of the sparger (Chisti, 1989).

There are two main classes of spargers, namely, static and dynamic. Perforated plates and pipes, porous plates and orifice spargers are examples of static spargers. Dynamic spargers are not widely used as they are complex to design and build and they require an external circulation of liquid along with additional pumping (Pillay, 2000).

Operating Flow Regimes

Flow configurations that are generated within the reactor are key in differentiating the ALR from other bioreactors. These flow configurations differ among the various sections of the ALR, hugely impacting its overall behaviour. In two-phase systems, the gas enters through a sparger but is distributed by the liquid recirculating through the downcomer (Merchuk & Siegel, 1988).

Three main flow regimes exist within ALRs which are dependent on bubble characteristics and reactor design.

- Homogenous/Bubbly Flow: Even distributions of bubbles that rise uniformly occur within this regime. A low flow rate is used and bubbles do not have much interaction (Abashar, et al., 1998).
- Heterogeneous/Churn Turbulent Flow: This regime results in coalescence of bubbles that produce larger bubbles which rise with a

higher rise velocity when compared to smaller bubbles. The flow rate in the heterogeneous regime is higher than that of homogeneous flow (Nigar Kantarci, 2004).

- Slug Flow: Mainly found in small diameter columns with high gas flow rates. As the gas velocity is increased, bubble slugs form due to stabilization of larger bubbles with the riser wall. As the bubbles move up the column, they coalesce with small sized bubbles producing elongated bubbles (Pillay, 2000). This type of flow regime is not favoured as the larger bubbles result in poor mass transfer however it is possible to convert slug flow into heterogeneous flow by either increasing the flow rate or increasing reactor diameter (Merchuk, and Siegel, 1988).

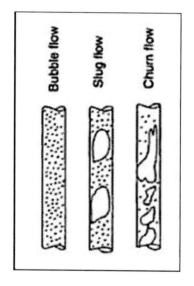

Figure 2. Flow Pattern in Two-Phase Flow (Coulson et al., 1999).

Hydrodynamic Properties

The performance of the ALR can be determined using hydrodynamic properties such as superficial gas velocity, mass transfer coefficients, superficial liquid circulation velocity and gas holdup. The latter two properties are considered to be the most important and are determined by the flow rate of gas injected into the reactor (Lu, Hwang and Cheng, 1994).

Gas Holdup

Gas holdup can be simply defined as the volume fraction of gas within the two-phase gas-liquid mixture.

$$\varepsilon_G = \frac{V_G}{V_G + V_L} \tag{1}$$

where:
    ε_G – gas holdup
    V_G – volume of gas (m^3)
    V_L – volume of liquid (m^3)

Gas holdup serves two significant functions. It gives an indication of the mean residence time of the gas in the liquid which is then used to ascertain the potential for mass transfer within the system. The larger the gas holdup, the larger the gas-liquid interfacial area hence larger mass transfer rates. The driving force for circulation is provided by the net hydrostatic pressure difference between the riser and downcomer due to the difference in gas holdups between the above-mentioned sections (Merchuk, and Gluz, 2000). Gas holdup increases with increasing superficial gas velocity. This increase is proportional in the homogeneous flow regime however in the heterogeneous regime; the effect is not as great due to the increased amount of larger bubbles that have a shorter residence time than smaller bubbles. In turn, the smaller bubbles contribute more to gas holdup and are preferred over larger bubbles. Factors that influence gas holdup include liquid properties, column dimensions, superficial gas velocity, operating temperature and pressure and sparger design (Pillay, 2000).

Many experiments have been performed on pilot-plant ALRs resulting in numerous correlations to calculate gas holdup, however, the following equation derived by Chisti (Chisti, 1989) will be used:

$$\varepsilon_G = \frac{\rho_L}{\rho_L - \rho_V} \cdot \frac{\Delta h_m}{\Delta z} \tag{2}$$

where:
    ε_G – gas holdup
    ρ_L – density of liquid (kg/m^3)
    ρ_V – density of gas (kg/m^3)
    Δh_m – differential heights (m)
    Δz – height difference between sampling points (m)

Superficial Liquid Circulation Velocity

As mentioned earlier, circulation of liquid occurs due to hydrostatic pressure differences between the riser and downcomer. Superficial liquid circulation velocity affects mixing time, mean residence time in the gas phase, interfacial area, mass transfer coefficients and gas holdup (Pillay, 2000).

Superficial liquid circulation velocity in the downcomer is calculated as follows:

$$U_{LD} = \frac{Q_L}{A_D} \tag{3}$$

where:

U_{LD} – downcomer superficial liquid circulation velocity (m/s)
Q_L – liquid flowrate (m^3/s)
A_D – cross-sectional area of the downcomer (m^2)

Cross-sectional area in the riser and downcomer are given by:

$$A_D = A_R = \frac{\pi d^2}{4} \tag{4}$$

where:

A_D - cross-sectional area of the downcomer (m^2)
A_R - cross-sectional area of the riser (m^2)
d – diameter (m)

Since $A_D = A_R$, superficial liquid circulation velocity can be determined by the continuity equation:

$$U_{LR} = U_{LD} \times \frac{A_D}{A_R} \tag{5}$$

where:

U_{LR} – riser superficial liquid circulation velocity (m/s)
A_D - cross-sectional area of the downcomer (m^2)
A_R - cross-sectional area of the riser (m^2)

Superficial Gas Velocity

Superficial Gas Velocity can be defined as the average velocity of the gas that has been sparged into the system.
It is measured with the aid of a rotameter and calculated as follows:

$$U_G = \frac{Q_G}{A_R} \tag{6}$$

where:
U_G – superficial gas velocity (m/s)
Q_G – gas flowrate (m³/s)
A_R - cross-sectional area of the riser (m²)

Gas Recirculation

Gas recirculation occurs when gas flowing up the riser becomes trapped in the liquid which causes it to circulate through the downcomer. This affects the flow configuration in the downcomer as well as the overall performance of the ALR (Chisti, 1989).

Liquid Level

The level of liquid in the ALR affects the disengagement of bubbles in the disengagement tank. A bubble will disengage or flow into the downcomer depending on free rising velocity of the bubble, velocity of liquid in the downcomer and residence time of the bubble in the disengagement tank (Merchuk & Gluz, 2000).

Increasing the liquid level decreases the amount of gas trapped, therefore there is less gas circulation in the downcomer which increases liquid velocity. In turn, the gas holdup in the riser and downcomer is reduced causing a higher rate of bubble disengagement (Pillay, 2000).

Overall Volumetric Mass Transfer Coefficient

Mass transfer in an airlift reactor is initiated by a concentration difference driving force and is modeled as a two-film model with resistance to mass transfer occurring primarily on the thin films situated near the interface. For sparingly soluble gases, resistance to mass transfer lies in the liquid film at the interface whilst gas phase resistance is negligible.

Equation, along with the dynamic absorption method, was used to calculate oxygen uptake rate. This equation indicates that the rate is directly proportional to the driving force of dissolved oxygen concentration difference.

$$\frac{d(DO)}{dt} = k_L a(DO^* - DO) \tag{7}$$

In order to calculate the overall volumetric mass transfer coefficient, equation is first integrated and then a graph of $ln\left(\frac{DO^* - DO}{DO^* - DO^0}\right)$ versus time is plotted. The gradient of the graph is equal to the overall volumetric mass transfer coefficient (Chisti, 1989).

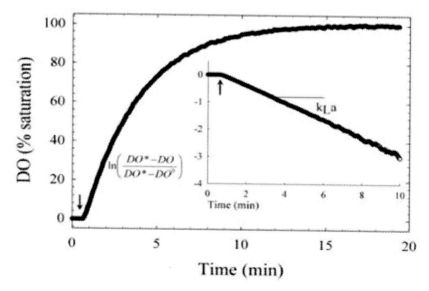

Figure 3. Dissolved Oxygen Saturation VS Time.

ADVANTAGES (*Jin, Yin and Lant, 2006*)

- Ease of construction with no moving parts
- Low maintenance
- Low energy requirements as compared to stirred tank reactors due to the absence of an agitator which consumes a lot of energy
- Rapid Mixing
- Good heat and mass transfer
- Good gas absorption efficiency
- Low shear rate
- Simple Scale Up
- Low shear force therefore suitable for microorganism growth as compared to bubble column or stirred tank reactors which result in a wide variety of shear forces.

DISADVANTAGES (*Jin, Yin and Lant, 2006*)

- Occurrence of foaming as there are no shafts or blades to break the foam and the use of anti-foam may affect growth of cells in bioreactors
- For highly viscous fluids, circulation velocity is reduced
- Air supply needs to be delivered at high pressures to achieve good agitation
- High capital costs
- Bubble coalescence in ALRs may cause the onset of plug flow which is highly undesirable
- Development of dead zones is a concern for ALRs

APPLICATIONS (*Jianping Wen, 2005*)

- Production of beer
- Production of citric acid
- Production of vinegar
- PRUTEEN Process
- Aerobic fermentation
- Culturing plant and animal cells
- Treatment of municipal and industrial waste water

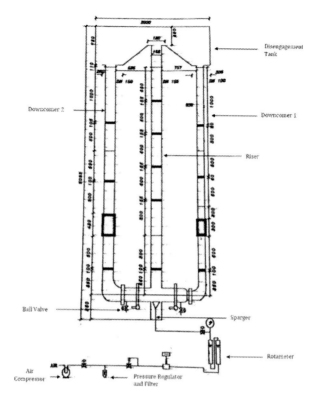

Figure 4. Schematic of External Loop Airlift Reactor.

EQUIPMENT AND MATERIALS

Materials

- Air
- Water

Equipment

- Three-column external loop airlift reactor
- Inverted Manometers
- Rotameter
- Dissolved Oxygen Sensor Probe

- Flowmetrix Flowmeter
- YSI Data Manager Software

Experimental Setup

- The reactor was initially flushed several times with water and disengagement tank scrubbed to remove any impurities or debris that may have built up in the reactor.
- The silicon tubing was then cut and cleaned extensively to remove build-up of algae that may have caused blockages during operation.
- To prevent flooding of manometers, stopcocks were installed just after each sampling point and just before each manometer.
- Gas and water lines were inspected for leaks.
- The dissolved oxygen probe was placed in the column at approximately a centre axial position to reduce any potential end effects that may occur. Thereafter, the probe was sealed with silicon tape to avoid water leaks from the column.

Figure 5. Installation of Stopcocks.

Figure 6. Placement and sealing of oxygen probe.

Experimental Procedure

1. Once all stopcocks and valves were closed, the ALR was filled with water to the desired level.
2. Pressure was set to 400kPa on the pressure regulator.
3. The rotatameter was set to the required value.
4. Stopcocks after each sampling point were opened.
5. Liquid circulation velocity was recorded from flowmeters for runs with downcomers. Four readings were taken and averaged for each run.
6. Stopcocks before each manometer were opened one-by-one to record manometer readings between sampling points.

The Dynamic Absorption Method was used in the following steps:

1. The dissolved oxygen (DO) concentration, temperature and pressure was recorded.
2. Real time acquisition data was captured by using the YSI Data Manager Software.
3. In order to avoid delays and lag times, the gas flowrate was set for the next run.

Mass Transfer Characteristics of an External Loop Airlift Reactor   145

4. The pressure regulator was set to 0kPa to ensure the water would not continue to be aerated between runs.

This procedure was repeated for five increasing gas velocities for each of the four configurations.

Configurations

1. Flow through only the riser
2. Flow through the riser and downcomer 2
3. Flow through the riser and downcomer 1
4. Flow through the riser and downcomer 1 and 2

Configuration 1: Riser (150mm diameter)

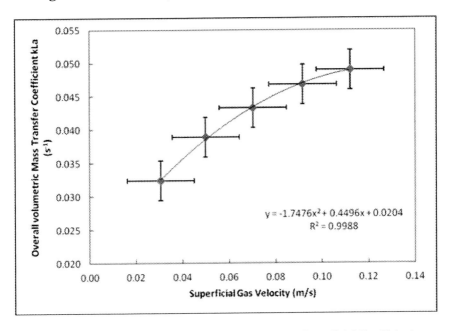

Figure 7. Overall Volumetric Mass Transfer Coefficient vs Superficial Gas Velocity-Riser.

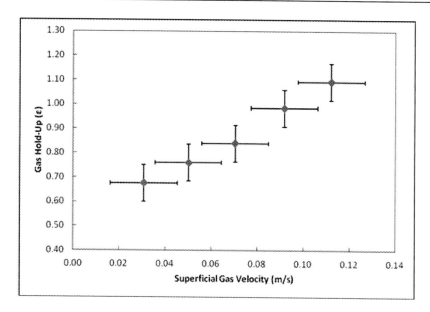

Figure 8. Gas Hold Up vs Superficial Gas Velocity-Riser.

Configuration 2: Riser and Downcomer 2 (150mm diameter)

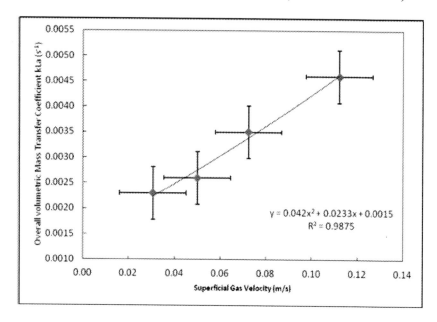

Figure 9. Overall Volumetric Mass Transfer Coefficient vs Superficial Gas Velocity-Riser and Downcomer 2.

Mass Transfer Characteristics of an External Loop Airlift Reactor 147

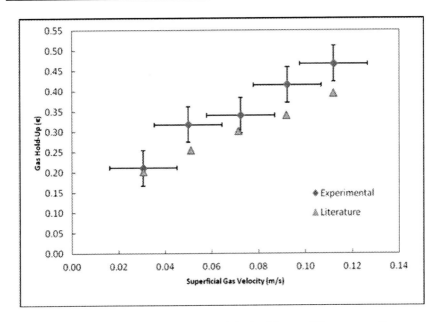

Figure 10. Gas Hold Up vs Superficial Gas Velocity-Riser and Downcomer 2.

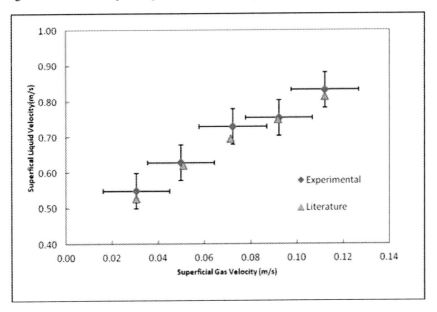

Figure 11. Superficial Liquid Velocity Vs Superficial Gas Velocity - Riser and Downcomer 2.

Configuration 3: Riser and Downcomer 1 (100mm diameter)

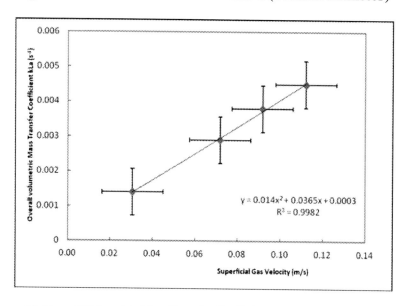

Figure 12. Overall Volumetric Mass Transfer Coefficient Vs Superficial Gas Velocity—Riser and Downcomer 1.

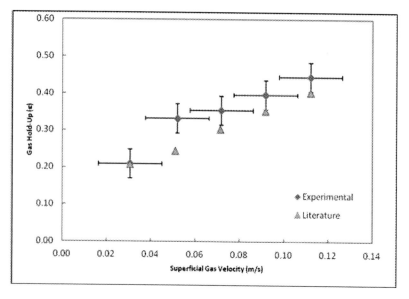

Figure 13. Gas Hold-Up Vs Superficial Gas Velocity-Riser and Downcomer 1.

Mass Transfer Characteristics of an External Loop Airlift Reactor   149

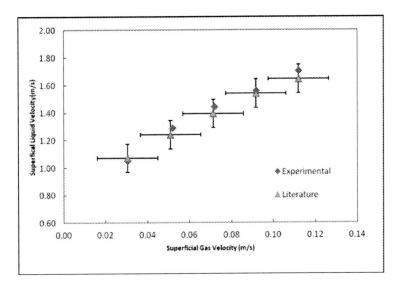

Figure 14. Superficial Liquid Vs Superficial Gas Velocity - Riser Downcomer 1.

Configuration 4: Riser, Downcomer 1 and Downcomer 2

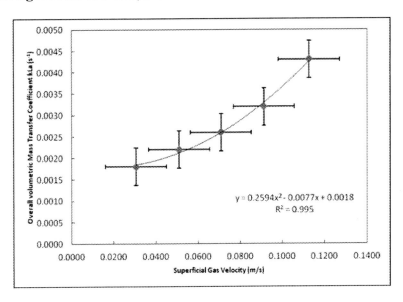

Figure 15. Overall Volumetric Mass Transfer Coefficient Vs Superficial Gas Velocity—Riser, Downcomer 1 and Downcomer 2.

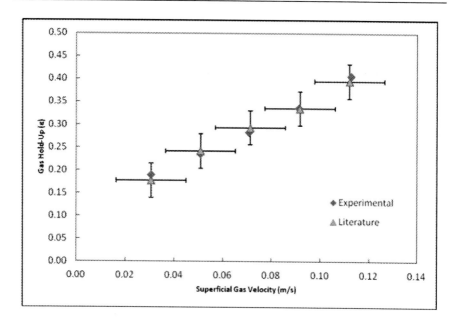

Figure 16. Gas Hold-Up Vs Superficial Gas Velocity- Riser, Downcomer 1 and Downcomer 2.

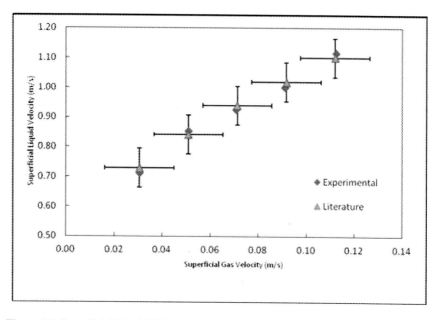

Figure 17. Superficial Liquid Velocity Vs Superficial Gas Velocity For Downcomer 1.

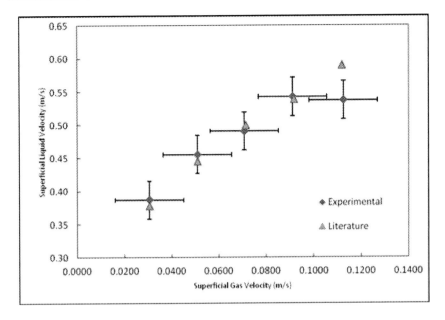

Figure 18. Superficial Liquid Velocity Vs Superficial Gas Velocity For Downcomer 2.

SUMMARY OF RESULTS

The results are summarized in the following Tables and Figures 7-18:

Table 1. Summary of superficial liquid and gas velocities, gas hold-ups and overall volumetric mass transfer coefficients for each reactor configuration

Run	Mean	Standard Deviation
Calibration		
ULD2 (m/s)	0.68	0.11
UGR (m/s)	0.07	0.03
εTotal	0.31	0.10
Riser		
UGR (m/s)	0.07	0.03
εTotal	0.87	0.17
kLa (s^{-1})	0.042080	0.006623
Riser & Downcomer 2		
ULD2 (m/s)	0.70	0.11
UGR (m/s)	0.07	0.03

Table 1. (Continued)

Run	Mean	Standard Deviation
εTotal(s^{-1})	0.35	0.10
kLa s^{-1}	0.003250	0.001034
Riser & Downcomer 1		
ULD1 (m/s)	1.41	0.25
UGR (m/s)	0.07	0.03
εTotal	0.35	0.09
kLa(s^{-1})	0.003150	0.001338
Riser, Downcomer 1 & Downcomer 2		
ULD1 (m/s)	0.92	0.15
ULD2 (m/s)	0.48	0.06
UGR (m/s)	0.07	0.03
εTotal	0.28	0.08
kLa(s^{-1})	0.002820	0.002820

Table 2. Arithmetic mean for the calculated uncertainty values for the overall mass transfer coefficients $K_la(S^{-1})$ for each configuration

Run	Uncertainty	Percent of calculated Value (%)
Riser	±0.002900	6.89
Riser & Downcomer 2	±0.000320	9.84
Riser & Downcomer 1	±0.000200	6.34
Riser, Downcomer 1 & Downcomer 2	±0.000142	5.03

DISCUSSION

The primary objective of this study was to determine the overall volumetric mass transfer coefficients of an external loop airlift reactor for a two phase air-water system. Initially, a few setbacks occurred which had to be overcome in order to proceed with the task. Blockages of algae were seen in silicon tubing throughout the reactor thus the tubing had to be cut and rid of algae and then reconnected or replaced with new silicon tubing. Flooding of the manometers was observed therefore stopcocks had to be installed just after each sampling point and just before each manometer. Lastly, the dissolved oxygen concentration was first ascertained by taking a sample of the mixture into a beaker by opening the stopcock after the sampling point. However, this method proved inaccurate as the oxygen in the water escaped into the atmosphere almost immediately after taking the sample. To combat this problem, the dissolved oxygen probe was placed through the sampling point and sealed with silicon tape to prevent leakages. This enabled an online testing method as the sensor from the probe was in direct contact with the fluid in the reactor, with data acquisition recorded instantaneously with the aid of the YSI Data Manager Software.

Gas Hold Ups

Five runs were performed for each configuration by varying gas flow rate which was then converted to superficial gas velocity using equation 6. Graphs of gas hold up versus superficial gas velocity were plotted as can be seen in Figures 8, 10, 13 and 16. In all four graphs, gas hold up was directly proportional to superficial gas velocity. In other words, as gas flow rate increased, gas hold up increased fairly linearly. This can be expected as increasing gas flow rate results in a larger amount of gas being sparged into the system hence higher gas hold ups. Greater gas flow rates are also indicative of higher liquid circulation velocity. Consequently, gas bubbles have less time to rise into the gas disengagement tank and are trapped in the downcomer causing a larger gas hold up. Conversely, when gas flow rate is low, liquid circulation velocity is reduced implying that gas bubbles have more time to rise in the disengagement tank, leading to a greater degree of gas disengagement, thus less bubbles move into the downcomer resulting in lower gas hold ups.

Configuration 1, which consisted of only the riser, had the highest gas hold up of 0.8701. This can be attributed to the position of the sparger as gas is sparged directly into the riser and remains there because the disengagement tank plays no role due to the water level ending at the top of the riser. All the gas is held up in the riser with a continuous flow of gas injection. Configuration 4, which was the riser, downcomer 1 and downcomer 2, had the lowest gas hold up of 0.2800. This is due to the fact that most of the bubbles disengage before travelling into the downcomers. Configuration 2 and 3 (riser and downcomer 2 and riser and downcomer 1) had gas hold ups of 0.3500 and 0.3472 respectively, which is in accordance as configuration 3 had a smaller downcomer diameter. Deviations from literature values on the graphs can be justified by discrepancies in gas flow rate. The valve used to set the flow rate on the rotameter was difficult to adjust, hence flow rate values varied slightly from those used in literature.

Liquid Circulation Velocity

During each run, liquid circulation flow rate was recorded however fluctuations in this reading occurred throughout the run and as such, an average of four values was used as the final value for calculation of liquid circulation velocity. The relationship between liquid circulation velocity and superficial gas flow rate were plotted on Figures 11, 14, 17 and 18. As superficial gas velocity increased, superficial liquid velocity increased as well. Pressure gradients are the driving force for liquid circulation which is a result of density differences of the fluid between the riser and downcomer sections. This, is turn, is controlled by the gas flow rate so increasing the gas flow rate increases the driving force and increases the liquid circulation velocity.

When comparing the liquid circulation velocity for downcomer 2 in configuration 3 and downcomer 1 in configuration 3, it was found that downcomer 1 had a higher liquid circulation velocity of 1.4089m/s attributed to it having a smaller diameter. Downcomer 1 and 2 in configuration 4 had lower liquid circulation velocities than those in configuration 2 and 3 as the volume of fluid now had to be split between both downcomers in configuration 4 whereas the fluid only had to flow through one downcomer in configurations 2 and 3 as can be seen in Table 1. The experimental results obtained correlate strongly to literature values which indicate accuracy of the runs.

Overall Volumetric Mass Transfer Coefficients

The specific bubble area was difficult to calculate, especially for small bubbles, and would require the use of photo analysis software which was beyond the scope of this study therefore the lumped parameter, k_La was considered. For sparingly soluble gases in water, such as oxygen, gas phase resistance to mass transfer is negligible. As superficial gas velocity increased, overall volumetric mass transfer coefficient increased (as can be seen in Figures 7, 9, 12 and 15) due to the fact that more gas was sparged into the system hence more gas was available for mass transfer.

The riser showed the greatest mass transfer coefficient of $0.0408s^{-1}$ which is justified as gas is sparged directly into the riser as previously mentioned. The gas is therefore able to transfer directly into the water phase without disengaging in the gas disengagement tank. The riser also has the smallest working volume which aids in mass transfer. One of the uses of downcomers in ALRs is to provide a greater degree of mixing however the runs containing downcomers showed low mass transfer coefficients. For future work, the liquid level in the gas disengagement tank should be decreased to ensure that fewer bubbles disengage allowing for more bubbles to travel to the downcomers for mass transfer.

Outliers were noticed on the data points in figures. These outliers were distinct at regions above 0.06m/s gas velocity which is indicative of a change in flow regime to more turbulent regime thus causing a greater degree of fluctuation.

Statistical analysis was performed on the overall volumetric mass transfer results. The greatest uncertainty calculated was 11% which is reasonable for this system which operated at conditions that were not 100% ideal.

From the results presented, it was found that configuration 1, only the riser, is the optimum configuration to use as it had the highest gas hold up, liquid circulation velocity and overall volumetric mass transfer coefficient. To obtain even better results, the gas flow rate should be increased beyond the values used in this study. This was not possible to do at present as the disengagement tank overflowed at higher gas flow rates and thus required a covering which was not available.

Further improvements can be made to this setup by allowing sparger position and hole diameter to be varied to determine the effect on mass transfer as well as varying liquid level to find the optimum level for greatest mass transfer.

The two phase air-water system used is a simple model for the external loop airlift reactor. This was investigated to gain knowledge on how the ALR works before expanding the system to a three phase system or scaling up for use in industry, in which case, variables such as diameter, volume, liquid level, density and viscosity, to name a few, would have to altered to suit the desired system.

Experimental runs could not be repeated to confirm accuracy of the results presented due to time constraints and wastage of water (each run uses over 500L of water). For use in industry, runs should be repeated and compared with these experimental values for repeatability and reproducibility.

CONCLUSION

- An increase in superficial gas velocity results in an increase in gas hold up as more gas is sparged into the system at higher flow rates.
- Configuration 1 produced the largest gas hold up of 0.8701 while configuration 4 had the lowest gas hold up 0f 0.2800.
- Greater gas disengagement occurs at low superficial gas velocities due to bubbles having more time to rise through the disengagement tank.
- Increasing superficial gas velocity increases superficial liquid velocity because at higher flow rates the pressure gradient driving force for liquid velocity is increased.
- The smaller diameter downcomer in configuration 3 had the larger superficial gas velocity when compared to configuration 2 and downcomer 1 and 2 in configuration 4 had the lowest liquid circulation velocities due to a split in flow path of the fluid.
- Superficial gas velocity is directly proportional to overall mass transfer coefficient.
- Configuration 1 had the greatest overall volumetric mass transfer coefficient as it had the lowest working volume and gas is sparged into the riser allowing direct transfer of mass.
- The highest uncertainty for overall volumetric transfer coefficient was 11%.
- Results obtained strongly correlated to literature values.
- Outliers in results were attributed to variations in flow rates used as well as a change in flow regime for velocities greater than 0.06m/s.
- In general, configuration 1 showed the best results in terms of gas hold up, liquid circulation velocity and overall volumetric mass transfer coefficient.

APPENDIX A: RAW DATA

Table A1. Calibration Data for Configuration 2 - Riser and Downcomer 2

RUN	PRESSURE IN (kPa)	LIQUID FLOW RATE: Downcomer 2 (L/s)	GAS FLOW RATE Rotameter reading	GAS FLOW RATE Actual (L/s)	ΔP_{riser} (mm H_2O) 4-5	5-6	6-7	7-8	$\Delta P_{downcomer\ 2}$ (mm H_2O) 1-2	2-3
1	400	9.0976	4	0.5405	25	15	26	24	4	26
2	400	10.746	7.8	0.8825	32	18	34	35	8	37
3	400	12.236	12	1.2605	37	24	41	41	9	49
4	400	13.247	16	1.6205	41	27	44	43	12	67
5	400	14.311	20	1.9805	46	33	51	47	14	65

Table A2. Recorded Experimental Calibration Data for Configuration 2 – Riser and Downcomer 2

RUN	PRESSURE IN (kPa)	LIQUID FLOW RATE: Downcomer 1 (L/s)	GAS FLOW RATE Rotameter Reading	GAS FLOW RATE Actual (L/s)	ΔP_{riser} (mm H_2O) 4-5	5-6	6-7	7-8	$\Delta P_{downcomer\ 2}$ (mm H_2O) 1-2	2-3
1	400	9.507	4.5	0.5855	21	14	25	28	4	29
2	400	10.846	8.1	0.9095	29	14	35	39	7	37
3	400	12.363	12.2	1.2785	31	26	35	49	7	52
4	400	13.174	15.9	1.6115	48	34	54	57	8	60
5	400	14.311	20	1.9805	58	34	63	42	11	65

Table A3. Calibration Results for Configuration 2 from Literature (Pillay, 2000)

RUN	Superficial liquid velocity in Downcomer 2: U_{LD2} (m/s)	Superficial gas velocity (riser) U_{GR} (m/s)	Gas Hold-up ε_g4-5	Gas Hold-up ε_g5-6	Gas Hold-up ε_g6-7	Gas Hold-up ε_g7-8	Gas Hold-up ε_g1-2	Gas Hold-up ε_g2-3	Gas Hold-up ε_g Riser	Gas Hold-up ε_g Downcomer 2	Total Gas Hold-up
1	0.5148	0.03058	0.02503	0.03003	0.05206	0.04805	0.002188	0.02603	0.1551	0.02821	0.1833
2	0.6080	0.04993	0.03203	0.03604	0.06808	0.07008	0.004376	0.03704	0.2062	0.04142	0.2476
3	0.6924	0.07133	0.03704	0.04805	0.08209	0.08209	0.004923	0.04905	0.2493	0.05398	0.3033
4	0.7496	0.09170	0.04105	0.05406	0.0881	0.0861	0.006565	0.06708	0.2693	0.07364	0.3429
5	0.8098	0.11200	0.04605	0.06608	0.1021	0.09411	0.007659	0.06507	0.3083	0.07273	0.3810

Table A4. Calibration Results from Experimental Data for Configuration 2

RUN	Superficial liquid velocity in Downcomer 2: U_{LD2} (m/s)	Superficial gas velocity (riser) U_{GR} (m/s)	Gas Hold-up ε_g4-5	Gas Hold-up ε_g5-6	Gas Hold-up ε_g6-7	Gas Hold-up ε_g7-8	Gas Hold-up ε_g1-2	Gas Hold-up ε_g2-3	Gas Hold-up ε_g Riser	Gas Hold-up ε_g Downcomer2	Total Gas Hold-up
1	0.5380	0.03313	0.02103	0.02803	0.05006	0.05606	0.00219	0.02903	0.1552	0.0312	0.1864
2	0.6138	0.05147	0.02903	0.02803	0.07008	0.07808	0.00383	0.03704	0.2052	0.0409	0.2461
3	0.6996	0.07235	0.03104	0.05205	0.07008	0.09810	0.00383	0.05206	0.2513	0.0559	0.3072
4	0.7455	0.09119	0.04806	0.06807	0.10812	0.11412	0.00438	0.06007	0.3384	0.0644	0.4028
5	0.8098	0.11207	0.05807	0.06807	0.12615	0.08409	0.00602	0.06508	0.3364	0.0711	0.4075

Table A5. Experimental Data for Configuration 1 – Riser

Run	Pressure In (kPa)	Gas Flowrate Rotameter Reading	Gas Flowrate Actual (L/s)	ΔPriser (mm H$_2$O) 4-5	5-6	6-7	7-8
1	400	4.00	0.5405	131	90	107	75
2	400	7.80	0.8825	165	102	101	94
3	400	11.80	1.2425	180	114	106	109
4	400	16.00	1.6205	155	140	83	191
5	400	20.00	1.9805	223	134	140	160

Table A6. Calculated Results for Configuration 1 – Riser

RUN	Superficial gas velocity (riser) U$_{GR}$ (m/s)	Gas Hold-up εg4-5	εg5-6	εg6-7	εg7-8	εgRiser
1	0.0306	0.1312	0.1802	0.2142	0.1502	0.6757
2	0.0499	0.1652	0.2042	0.2022	0.1882	0.7598
3	0.0703	0.1802	0.2282	0.2122	0.2182	0.8389
4	0.0917	0.1552	0.2803	0.1662	0.3824	0.9841
5	0.1121	0.2233	0.2683	0.2803	0.3203	1.0922

Table A7. Experimental Data for Configuration 2 - Riser and Downcomer 2

Run	Pressure In (kPa)	Liquid Flowrate: Downcomer 2 (L/s)	Gas Flowrate Rotameter Reading	Gas Flowrate Actual (L/s)	ΔPriser (mm H$_2$O) 4-5	5-6	6-7	7-8	Δpdowncomer 2 (mm H$_2$O) 1-2	2-3
1	400	9.6961	4.00	0.5405	35	13	31	30	8	23
2	400	11.0940	7.80	0.8825	37	35	38	41	11	46
3	400	12.8890	12.20	1.2785	45	18	46	47	9	68
4	400	13.3270	16.10	1.6295	55	39	54	57	10	54
5	400	14.7100	20.00	1.9805	54	48	60	58	10	75

Table A8. Calculated results for Congifuration 2 - Riser and Downcomer 2

RUN	Superficial liquid velocity Downcomer 2: U_{LD2} (m/s)	Superficial gas velocity (riser) U_{GR} (m/s)	Gas Hold-up ε_g4-5	ε_g5-6	ε_g6-7	ε_g7-8	ε_g1-2	ε_g2-3	ε_gRiser	ε_gDowncomer 2	Total
1	0.5487	0.0306	0.0350	0.0260	0.0621	0.0601	0.0044	0.0230	0.1832	0.0274	0.2106
2	0.6278	0.0499	0.0370	0.0701	0.0761	0.0821	0.0060	0.0461	0.2653	0.0521	0.3174
3	0.7294	0.0723	0.0451	0.0360	0.0921	0.0941	0.0049	0.0681	0.2673	0.0730	0.3403
4	0.7542	0.0922	0.0551	0.0781	0.1081	0.1141	0.0055	0.0541	0.3554	0.0595	0.4149
5	0.8324	0.1121	0.0541	0.0961	0.1201	0.1161	0.0055	0.0751	0.3864	0.0806	0.4670

Table A9. Experimental Data for Configuration 3 - Riser and Downcomer 1

Run	Pressure In (kPa)	Liquid Flowrate: Downcomer 1 (L/s)	Gas Flowrate Rotameter Reading	Actual (L/s)	ΔP_{riser} (mm H$_2$O) 4-5	5-6	6-7	7-8	$\Delta p_{downcomer 1}$ (mm H$_2$O) 9-10	10-11
1	400	8.2488	4.00	0.5405	39	22	30	16	2	33
2	400	10.1138	8.20	0.9185	45	35	41	41	5	50
3	400	11.3378	12.10	1.2695	45	28	50	47	6	55
4	400	12.2535	16.00	1.6205	60	32	48	52	3	70
5	400	13.3753	20.00	1.9805	60	45	52	54	4	80

Table A10. Calculated Results for Configuration 3 - Riser and Downcomer 1

RUN	Superficial Liquid Velocity In Downcomer 1: U_{LD1} (m/s)	Superficial Gas Velocity (Riser) U_{GR} (m/s)	Gas Hold-up ε_g 4-5	ε_g 5-6	ε_g 6-7	ε_g 7-8	ε_g 9-10	ε_g 10-11	ε_gRiser	ε_gDowncomer1	Total
1	1.0503	0.0306	0.0390	0.0440	0.0601	0.0320	0.0011	0.0330	0.1752	0.0341	0.2093
2	1.2877	0.0520	0.0451	0.0701	0.0821	0.0821	0.0027	0.0501	0.2793	0.0528	0.3321
3	1.4436	0.0718	0.0451	0.0561	0.1001	0.0941	0.0033	0.0551	0.2953	0.0583	0.3537
4	1.5602	0.0917	0.0601	0.0641	0.0961	0.1041	0.0016	0.0701	0.3244	0.0717	0.3961
5	1.7030	0.1121	0.0601	0.0901	0.1041	0.1081	0.0022	0.0801	0.3624	0.0823	0.4447

Table A11. Experimental Data for Configuration4 - Riser, Downcomer 1 and Downcomer 2

Run	Pressure In (kPa)	Liquid Flowrate: Downcomer 1 (L/s)	Liquid Flowrate: Downcomer 2 (L/s)	Gas Flowrate Rotameter Reading	Gas Flowrate Actual (L/s)	ΔPriser (mm H2O) 4-5	5-6	6-7	7-8	Δpdowncomer 2 (mm H2O) 1-2	2-3	Δpdowncomer 1 (mm H2O) 9-10	10-11
1	400	5.5922	6.8365	4	0.5405	27	17	20	25	3	20	3	15
2	400	6.6913	8.0469	8	0.9005	35	22	27	30	2	25	2	15
3	400	7.2668	8.672	11.9	1.2515	40	26	35	38	4	25	2	16
4	400	7.857	9.5899	15.9	1.6115	49	34	40	39	2	32	3	26
5	400	8.7644	9.4967	20.1	1.9895	56	37	48	52	3	36	4	36

Table A12. Calculated Results for Configuration 4 - Riser, Downcomer 1 and Downcomer 2

RUN	Superficial Liquid Velocity In Downcomer 1: U_{LD1} (m/s)	Superficial Liquid Velocity In Downcomer 2: U_{LD2} (m/s)	Superficial Gas Velocity (riser) U_{GR} (m/s)	Gas Hold-up $\varepsilon_g 4\text{-}5$	$\varepsilon_g 5\text{-}6$	$\varepsilon_g 6\text{-}7$	$\varepsilon_g 7\text{-}8$	$\varepsilon_g 1\text{-}2$	$\varepsilon_g 2\text{-}3$	$\varepsilon_g 9\text{-}10$	$\varepsilon_g 10\text{-}11$
1	0.7120	0.3869	0.0306	0.0270	0.0340	0.0400	0.0501	0.0016	0.0200	0.0016	0.0150
2	0.8520	0.4554	0.0510	0.0350	0.0440	0.0541	0.0601	0.0011	0.0250	0.0011	0.0150
3	0.9252	0.4907	0.0708	0.0400	0.0521	0.0701	0.0761	0.0022	0.0250	0.0011	0.0160
4	1.0004	0.5427	0.0912	0.0491	0.0681	0.0801	0.0781	0.0011	0.0320	0.0016	0.0260
5	1.1159	0.5374	0.1126	0.0561	0.0741	0.0961	0.1041	0.0016	0.0360	0.0022	0.0360

Table A13. Gas Hold Ups for Configuration 4 - Riser, Downcomer 1 and Downcomer 2

Gas Hold-up			
ε_gRiser	ε_gDowncomer2	ε_gDowncomer1	Total
0.1512	0.0217	0.0167	0.1895
0.1932	0.0261	0.0161	0.2354
0.2383	0.0272	0.0171	0.2826
0.2753	0.0331	0.0277	0.3361
0.3304	0.0377	0.0382	0.4063

Table A14. Operating Conditions

Configuration	T(°C)	P(kPa)	DO_o (mg/L)	DO*(mg/L)
Riser				
Run 1	24.60	100.01	7.53	8.18
Run 2	24.30	99.97	7.11	8.23
Run 3	24.30	99.96	7.30	8.23
Run 4	24.40	99.93	6.84	8.21
Run 5	24.4	99.90	6.81	8.21
Riser & Downcomer 2				
Run 1	24.40	99.86	6.93	8.20
Run 2	24.50	99.91	6.79	8.19
Run 3	24.60	99.95	6.75	8.17
Run 4	23.20	100.01	7.54	8.19
Run 5	23.80	99.84	7.15	8.19
Riser & Downcomer 1				
Run 1	23.90	100.72	7.30	8.29
Run 2	24.00	100.18	7.76	8.26
Run 3	23.90	100.09	7.02	8.29
Run 4	23.90	100.01	7.14	8.22
Run 5	24.10	99.91	6.96	8.25
Riser, Downcomer 1 & Downcomer 2				
Run 1	24.10	99.81	6.96	8.26
Run 2	24.00	99.77	6.95	8.19
Run 3	23.80	99.81	6.87	8.29
Run 4	24.30	99.40	6.86	8.22
Run 5	24.20	99.32	6.87	8.23

APPENDIX B: CALIBRATION CHART

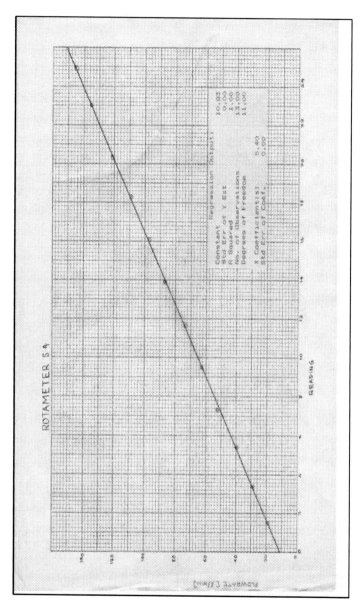

Figure B1. Calibration Chart for Low Flow Rates Fisher Controls KDG2000 Rotameter.

APPENDIX C: SAMPLE CALCULATIONS

Sample calculations were performed on configuration 4 which consisted of the Riser, Downcomer 1 and Downcomer 2.

The physical properties of air and water to be used in the following calculations were obtained from tables of Physical Property data at 24°C and 101.325kPa (Green & Perry, 2008).

Fluid Physical Properties

ρ_l (density of water) = 997.04kg/m^3
ρ_v (density of air) = 1.204kg/m^3

Reactor Geometry

Cross-Sectional Area
The cross-sectional area for the riser and both downcomers were calculated using equation 4.

$$A_r = \frac{\pi d_r^2}{4} = \frac{\pi (0.15)^2}{4} = 0.0177 \: m^2$$

$$A_{d1} = \frac{\pi d_{d1}^2}{4} = \frac{\pi (0.1)^2}{4} = 0.0079 \: m^2$$

$$A_{d2} = \frac{\pi d_{d2}^2}{4} = \frac{\pi (0.15)^2}{4} = 0.0177 \: m^2$$

Superficial Gas Velocity

The equation below was derived from a trend line fitted onto the calibration chart in Appendix B to convert rotameter readings.

$$Y = \frac{(5.4X + 10.83)}{60}$$

where:

X = Rotameter Reading
Y = Actual Reading (L/s)

For run 1:

$$Y = \frac{(5.4(4) + 10.83)}{60} = \frac{0.5405 L}{s} = Q_G$$

The superficial gas velocity can then be determined from equation 6.

$$U_{GR} = \frac{Q}{A_r} = \frac{0.5405}{0.01767} = 30.58 \ L/s = 0.0306 \ m/s$$

Superficial Liquid Velocity

Flowmetrix flow meters were used to measure the volumetric flow rate of the liquid in the downcomers. Due to fluctuations, an average of four readings was used and using equation 3, superficial liquid velocity was calculated.

$$U_{LD1} = \frac{Q}{A_{D1}} = \frac{5.5922}{0.0079} = 712.0210 \ L/s = 0.7120 \ m/s$$

$$U_{LD2} = \frac{Q}{A_{D2}} = \frac{6.8365}{0.0177} = 386.8668 \ L/s = 0.3869 \ m/s$$

Gas Hold Up

Manometer readings were used to determine gas hold-up between each pair of sampling points in the riser and downcomer, along with physical property data and height difference between sampling points.

From equation 2 for points 4-5:

$$\varepsilon_G = \frac{\rho_L}{\rho_L - \rho_v} \cdot \frac{\Delta h_m}{\Delta z} = \frac{997.04}{997.04 - 1.204} \cdot \frac{(27)}{1000.01} = 0.0270$$

The procedure was repeated for the remaining pairs of sampling points and then summed up to give a total gas hold up of $\varepsilon_G = 0.1895$.

Overall Volumetric Mass Transfer Coefficient

The overall volumetric mass transfer coefficient is obtained from the gradient of the $\ln \left(\frac{DO^* - DO}{DO^* - DO^0}\right)$ versus time plot with a trend line fitted shown in

Mass Transfer Characteristics of an External Loop Airlift Reactor 167

Figure C2. DO* and DO^0 are specific to each run with DO^0 being the initial concentration and DO* the equilibrium concentration at the specific temperature and pressure.

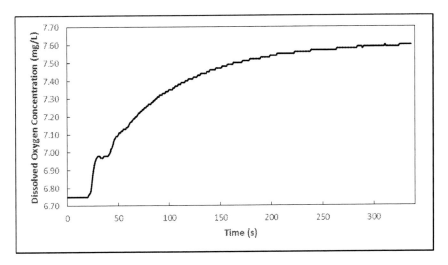

Figure C1. Dissolved oxygen concentration profile vs time for RUN 3 of the Riser and Downcomer 2 configuration.

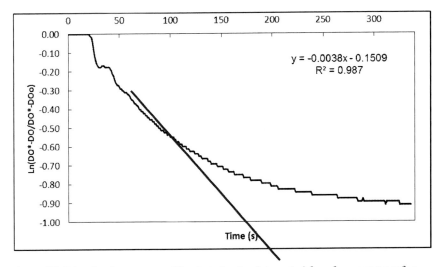

Figure C2. Dissolved oxygen profile plotted using integrated for of oxygen transfer equation used to obtain $kLa(s^{-1})$.

The absolute value of the gradient of the tangent to the linear part of the curve shown by the red line in Figure C2 is equal to the overall volumetric mass transfer coefficient. of run 3, of the riser and downcomer 2 configuration. The k_La value was found to be $0.0038s^{-1}$.

The uncertainty of the volumetric mass transfer coefficient is given by the gradient of $ln\left(\frac{DO^*-DO}{DO^*-DO^0}\right) \pm \frac{d}{d(DO)}\left(ln\left(\frac{DO^*-DO}{DO^*-DO^0}\right)\right)\Delta q$ Versus time graph. The second term accounting for the uncertainty in measurement, where Δq is the uncertainty in the dissolved oxygen probe (0.04mg/L) as specified by the manufacture. A similar calculation approach to find the overall volumetric mass transfer coefficient was followed. The difference in the results was taken to be the uncertainty in the calculated value. It was found that the lower and upper uncertainty values were equal.

REFERENCES

Abashar, M. E., Narsingh, U., Rouillard, A. E. and Judd, R. (1998). *Hydrodynamic Modelling of an External Loop Airlift Reactor* (Vol. 37). Ind. Eng. Chem. Res.

Chisti, M. (1989). *Airlift Bioreactors*. New York: Elsevier Science-Publishers LTD.

Coulson, J.M., Richardson, J.F., Backhurst, J.R. and Harker J.H. (1999). *Chemical Engineering Vol. I: Fluid Flow, Heat Transfer and Mass Transfer*, Sixth Edition.

Green D.W. and Perry R.H. (2008). *Perry's Chemical Engineers' Handbook*, Eighth Edition.

Jianping Wen, X. J. (2005). *Characteristics Of Three-Phase Internal Loop Airlift Bioreactors With Complete Gas Recirculation For Non-Newtonian Fluids..*

Jin, B., Yin, P. and Lant, P. (2006). *Hydrodynamics And Mass Transfer Coefficient In Three-Phase Air-Lift Reactors Containing Activated Sludge.*

Lu, W. J., Hwang, S. J. and Cheng, C. M. (1994). *Liquid Velocity And Gas Holdup In A Three Phase System.*

Merchuk, J. C. and Gluz, M. (2000). *Biorectors, Air-Lift Reactors.*

Merchuk, J. C. and Siegel, M. H. (1988). *Air-Lift Reactors In Chemical And Biological Technology.*

Nigar Kantarci, F. B. (2004). *Bubble Column Reactors.*

Pillay, C. (2000). *Hydrodynamics Of Airlift Systems*. Durban: University of Durban Westville.

Zhonghuo, D., Tiefeng, W., Nian, Z. and Zhanwen, W. (2010). *Gas Holdup, Bubble Behavior And Mass Transfer In A 5m High Internal-Loop Airlift Reactor With Non-Newtonian Fluid.*

In: Advances in Chemistry Research. Volume 35 ISBN: 978-1-53610-734-0
Editor: James C. Taylor © 2017 Nova Science Publishers, Inc.

Chapter 9

APPLICATIONS OF ISOLATION AND STRUCTURE ELUCIDATION OF SECONDARY METABOLITES IN NATURAL PRODUCT CHEMISTRY LABORATORY

Aliefman Hakim[1] and A. Wahab Jufri[2]
[1]Study Program of Chemistry Education, University of Mataram,
Mataram-Lombok, Indonesia
[2]Study Program of Biology Education, University of Mataram,
Mataram-Lombok, Indonesia

ABSTRACT

Natural products chemistry examines secondary metabolites contained in an organism, so it is strongly associated with pharmaceuticals, cosmetics, and pesticides. The chemical study of natural product based on experimental development and applications demanding high standards of laboratory activities. The laboratory activities involve the isolation of secondary metabolites from plants. The same secondary metabolites from a plant species can be isolated in a various ways, so there is no standard procedure to isolate the secondary metabolites of a plant species. These conditions can be used to train high-level thinking skills of learners. In natural product chemitry laboratory, learners can be given responsibility to undertake project to isolates the secondary metabolites from a variety of plant species. This laboratory works provide opportunities for students to design their own activities to isolate the secondary metabolites from medicinal plants. Students are exposed to skills as extraction, fractionation, purification, and structural elucidation of secondary metabolites. These laboratory activities can be useful for students at the third-year undergraduate level from many different disciplines including chemistry education, chemistry, pharmacy, and medicine.

Keywords: natural products, secondary metabolites, isolation, structure elucidation, laboratory

INTRODUCTION

Humanity is dependent on nature. People can obtain food, medicines, building materials, and other resources from the nature. Plants, animals, and microorganisms are sources of secondary metabolites diversity. Diversity of secondary metabolites requires isolation biodiversity through extraction, fractioation, purification, and structure elucidation of secondary metabolites. Natural products chemistry course examines secondary metabolites contained in an organism [1]. Total of 250.000 species of higher plants grow around the world [2]. About 85.000 species are grown in Latin America, 35.000 species in Africa, and at least 50.000 in Asia [2]. Less than 10% of the world's higher plants has studied the content of secondary metabolites [3]. These natural wealth has a great potential in advancing the natural products chemistry.

Studies of secondary metabolites through laboratory activities in natural product chemistry course useful for third or fourth year undergraduate students who have a basic understanding of the chromatographic and spectroscopic techniques used in the identification of natural compounds. There are at least two important points in natural product chemistry laboratory (i) procedures to isolate secondary metabolites from an organism, and (ii) elucidation structures of secondary metabolites.

PROCEDURES TO ISOLATE SECONDARY METABOLITES FROM AN ORGANISM

Of the hundreds of secondary metabolites that can be isolated from plants, many of them show interesting biological activities such as cytotoxicity [4-6], antimalarial activity [7-9], antiviral activity [10-12], antifungal activity [13-15], and antimicrobial activity [16, 17]. These biological activities can be used to guide students to the active pure compound isolation. Various bioactivities show potential for the lead compound to be useful for industrial drug or pesticide industries. Some examples of compounds isolated from plants such as artelastisin (*1*) and artelastin (*2*) isolated from *Artocarpus scortechinii* are shown in Figure 1 [18]. These compounds are flavonoid derivatives.

Isolation of secondary metabolites from various plant species provides an opportunity for students to find evidence to support the concept polar compounds will be soluble in polar solvents and nonpolar compounds will dissolve in nonpolar solvents, or to connect new concepts with the students knowledge to rationalize various phenomena such as the various properties of plants that can be used by humans for treatment. Efficacy of these plants relate to levels of the chemical content in the plants. These activities enhance the meaningful learning in class. Isolation secondary metabolites from plants through the process of extraction, fractionation, purification, and characterization has resulted in many compounds such as artoindonesianin A, B, C [19, 20]. General procedure to isolate secondary metabolites from plants according the following scheme (Figure 2) [21].

Although there is a general procedure for isolating secondary metabolites from an organism, but the same secondary metabolite from a plant species can be isolated in various ways. Futhermore there is no standard procedure to isolate the secondary metabolites of a plant species. These conditions can be used to train the creative thinking skills. The examples of innovative procedures of isolation of pinostrobin from *Kaemferia pandurata* compared with the literature presented in Figure 3 [21]. The active constituent, pinostrobin, is already known in the rhizomes of *K. pandurata* base on the literature [22, 23]. Futhermore students were asked to develop a procedure base on the literature and the results of each stage of isolation process. Procedure of isolation of pinostrobin from *Kaemferia pandurata* were developed by students which were different from the literature. Incompatibility of the results of laboratory implementation with the literature led the students to find their own procedures as seen in Figure 3. Differences in the procedures appear on purification stage and used eluent. Students used recrystallization with n-hexane as a solvent while literature used HPLC [23]. It occurs because the result of fractionation stage showed a crystalin fraction that may be purified through the recrystallization process. These laboratory activities provide opportunities for students to develop their crative thinking skills.

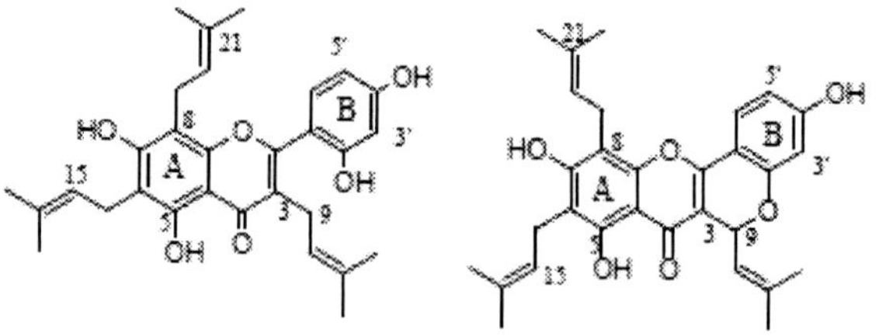

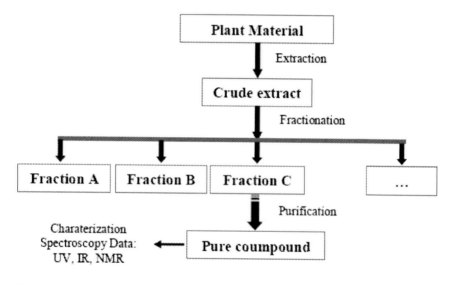

Figure 1. Secondary metabolites from *Artocarpus scortechinii*.

Figure 2. General stages of isolation of secondary metabolites.

ELUCIDATION STRUCTURES OF SECONDARY METABOLITES

Structure elucidation of secondary metabolites based on spectroscopic data. UV (Ultraviolet) spectrum is useful for determining the conjugated double bonds in the secondary metabolites. IR (Infra Red) spectrum is useful for identifying functional groups of secondary metabolites. NMR spectrum (Nuclear Magnetic Reconance) is useful to determine the arrangement of atoms of H and C in secondary metabolites. For examples of structural elucidation of pinostrobin from *K. pandurata* are presented [21]. UV spectra were measured with Beckman DU-7000 UV spectrophotometer and IR spectra were recorded on a Perkin Elmer FTIR spectrophotometer. H- and C-NMR spectra were taken on a Agilent 500 MHz spectrometer operating at 500.1 MHz (^{1}H) and 125.8 MHz (^{13}C).

UV spectrum in Figure 4 [λ_{maks} (log ε) 290 (0.99); and 239 (0.302) nm] was consistent with the presence of a flavanone structure.

The IR spectrum in Figure 5 showed absorptions for hydroxyl (3469 cm^{-1}), aliphatic C-H (2972 and 2910 cm^{-1}), and conjugated carbonyl (1643 cm^{-1}).

The ^{1}H-NMR spectrum in Figure 6 included signals for five proton aromatic (B ring) (δppm 7.365, *m*) and signals for two proton aromatic at C-6 and C-8 (δppm 6.050, 2H, *d, J* = 6 Hz). Proton at C-2 can be seen at (δppm) 5.356 (1H, *dd, J* = 13.5; 3 Hz), and signals at at (δppm) 3.072 (1 H, *dd, J* = 13; 4 Hz) and 2.799 (H, *dd, J* = 17; 2 Hz) showed two proton at C-3. Methoxy at C-7 was showed at δ 3.77 (3H, *s*), similar to the arrangement found for a related compound, pinostrobin (*1*). Supporting evidence for the structure assigned to pinostrobin (*1*) came from comparison of the ^{13}C-NMR spectrum in Figure 7. Three carbon Sp^3, twelve proton aromatic, and one carbon carbonil can be found similar to pinostrobin (*1*).

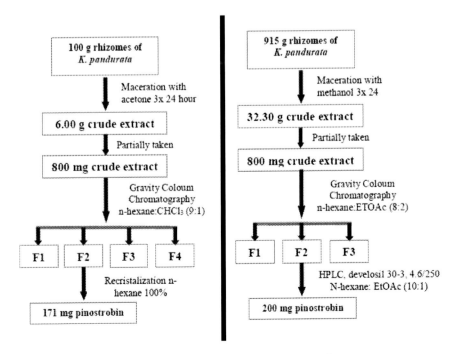

Figure 3. Procedures to isolate pinostrobin from *Kaemferia pandurata*.

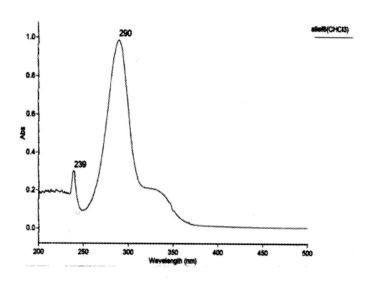

Figure 4. UV spectrum of pinostrobin from *K. pandurata*.

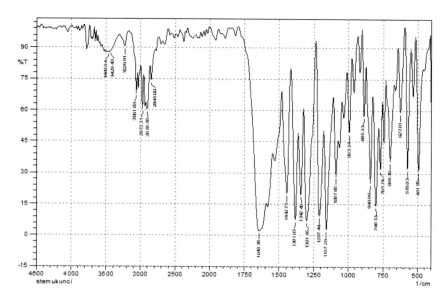

Figure 5. IR spectrum of pinostrobin from *K. pandurata*.

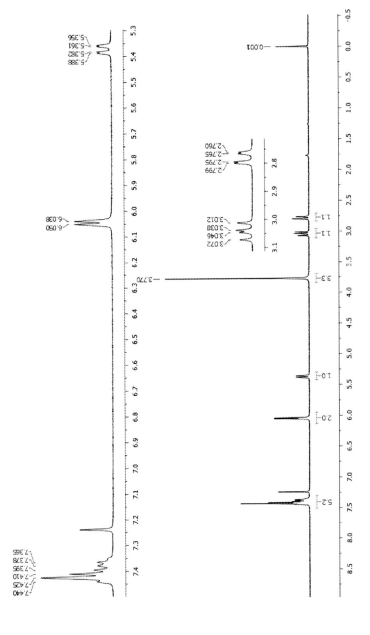

Figure 6. ¹H-NMR spectrum of pinostrobin from *K. pandurata*.

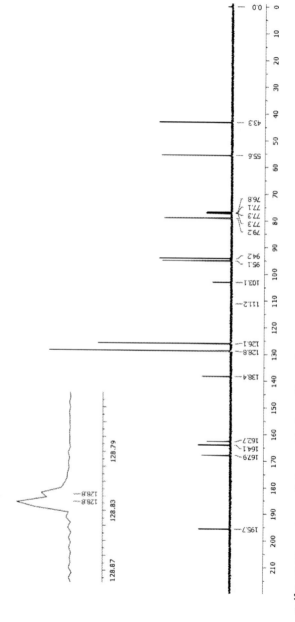

Figure 7. ^{13}C-NMR spectrum of pinostrobin from *K. pandurata*.

CONCLUSION

One type of laboratory that can be applied in natural product chemistry course is project-based laboratory. Student can be given a project to isolate secondary metabolites from plant species. The sample selection of plant species for natural product laboratory should be considered the level of difficulty of secondary metabolites isolation to avoid any impact on the student frustration due to failure in the laboratory activities.

REFERENCES

[1] Hakim, A., Liliasari, Kadarohman, A. Student understanding of Natural Product Concept of Primary and Secondary Metabolites Using CRI Modified. *International Online Journal of Educational Sciences* 2012, 4 (3), 544-553.
[2] Centres of The Plant Diversity Americas. http://botany.si.edu/projects/cpd/introduction.htm (accessed 02-06-2016).
[3] Dias, D.A., Urban, S., Roessner, U. A Historical Overview of Natural Products in Drug Discovery. *Metabolites* 2012, 2(1), 303-336.
[4] Pham, C., Hartmann, R., Müller, W. E. G., Voogd, N., Lai, D., Proksch, P. Aaptamine Derivatives from the Indonesian Sponge *Aaptos suberitoides*. *J. Nat. Prod.* 2013, 76(1), 103-106.
[5] Niemann, H., Lin, W., Müller, W. E. G., Kubbutat, M., Lai, D., Proksch P. Trimeric Hemibastadin Congener from the Marine Sponge *Ianthella basta*. *J. Nat. Prod.* 2013, 76(1), 121-125.
[6] Pham, C., Weber, H., Hartmann, R., Wray, V., Lin, W., Lai, D., Proksch, P. New Cytotoxic 1,2,4-Thiadiazole Alkaloids from the Ascidian *Polycarpa aurata*. *Org. Lett.* 2013, 15(9), 2230-2233.
[7] Blair, L. M., Sperry, J. Natural Products Containing a Nitrogen–Nitrogen Bond. *J. Nat. Prod.* 2013, 76(4), 794-812.
[8] Hakim, A. and Jufri, A. W. "Aktivitas Antimalaria dan Analisis Metabolit Sekunder Kayu dan Kulit Batang *Artocarpus odoratissimus* Blanco. (Moraceae)." *Jurnal Bahan Alam Indonesia* 2011, 7(6), 302-305.
[9] Kardono, L. B. S., Angerhofer, C. K., Tsauri, S., Padmawinata, K., Pezzuto, John M.A. Kinghorn, D. Cytotoxic and Antimalarial

Constituents of the Roots of Eurycoma longifolia. *J. Nat. Prod.* 1991, 54 (5), 1360-1367.
[10] Elfahmi, Batterman, S., Koulman, A., Hackl, T., Bos, R., Kayser, O., Woerdenbag, H. J., Quax, W. J. Lignans from Cell Suspension Cultures of *Phyllanthus niruri*, an Indonesian Medicinal Plant. *J. Nat. Prod.* 2006, 69(1), 55-58.
[11] Angawi, R. F., Calcinai, B., Cerrano, C., Dien, H. A., Fattorusso, E., Scala, F., and Scafati, O. T. Dehydroconicasterol and Aurantoic Acid, a Chlorinated Polyene Derivative, from the Indonesian Sponge *Theonella swinhoei*. *J. Nat. Prod.* 2009, 72(12), 2195-2198.
[12] Zainuddin, E. N., Mentel, R., Wray, V., Jansen, R., Nimtz, M., Lalk, M., Mundt, S. Cyclic Depsipeptides, Ichthyopeptins A and B, from *Microcystis ichthyoblabe*. *J. Nat. Prod.*, 2007, 70(7), 1084-1088.
[13] Handayani, D., Edrada, R. A., Proksch, P., Wray, V., Witte, L., Soest, R. W. M. V., Kunzmann, A., Soedarsono. Four New Bioactive Polybrominated Diphenyl Ethers of the Sponge *Dysidea herbacea* from West Sumatra, Indonesia. *J. Nat. Prod.*, 1997, 60(12), 1313-1316.
[14] Gu, J. Q., Graf, T. N., Lee, D., Chai, H. B., Mi, Q., Kardono, L. B. S., Setyowati, F. M., Ismail, R., Riswan, S., Farnsworth, N. R., Cordell, G. A., Pezzuto, J.M., Swanson, S. M., Kroll, D. J., Falkinham, J. O., Wall, M. E., Wani, M. C., Kinghorn, A. D., Oberlies N. H. Cytotoxic and Antimicrobial Constituents of the Bark of *Diospyros maritima* Collected in Two Geographical Locations in Indonesia. *J. Nat. Prod.*, 2004, 67(7), 1156-1161.
[15] Kosela, S., Hu, L. H., Rachmatia, T., Hanafi, M., Sim, K. Y. Dulxanthones F-H, Three New Pyranoxanthones from *Garcinia dulcis*. *J. Nat. Prod.* 2000, 63(3), 406-407.
[16] Nakazawa, T., Xu, J., Nishikawa, T., Oda, T., Fujita, A., Ukai, K., Mangindaan, R. E. P., Rotinsulu, H., Kobayashi, H., Namikoshi, M. Lissoclibadins 4-7, Polysulfur Aromatic Alkaloids from the Indonesian Ascidian *Lissoclinum* cf. *badium*. *J. Nat. Prod.* 2007, 70(3), 439-442.
[17] Yasman, Edrada, R. A., Wray, V., Proksch, P. New 9-Thiocyanatopupukeanane Sesquiterpenes from the Nudibranch *Phyllidia varicosa* and Its Sponge-Prey *Axinyssa aculeata*. *J. Nat. Prod.* 2003, 66 (11), 1512-1514.
[18] Hakim, A. A Prenylated Flavone from The Heartwood of *Artocarpus Scortechinii* King (Moraceae). *Indonesian Journal of Chemistry* 2009, 9 (1), 146-150.

[19] Hakim, E. H., Fahriyati, A., Kau, M. S., Achmad, S. A., Makmur, L., Ghisalberti, E. L., and Nomura, T. Artoindonesianins A and B, two new prenylated flavones from the root bark of Artocarpus champeden. *J. Nat. Prod.* 1999, 62(4), 613-615.

[20] Makmur, L., Syamsurizal, Tukiran, Achmad, S. A., Aimi, N., Hakim, E. H., Kitajima, M., and Takayama H. Artoindonesianin C, A New Xanthone Derivative from Artocarpus teysmanii. *J. Nat. Prod.* 2000, 63 (2), 243-244.

[21] Hakim, A., Liliasari, Kadarohman, A., Syah, Y.M. (2016). Making a Natural Product Chemistry Course Meaningful with a Mini Project Laboratory. *J. Chem. Educ.* 2016, 93(1), 193-196.

[22] Shindo, K., Kato, M., Kinoshita, A., Kobayashi, A., and Koike, Y. Analysis of antioxidant activities contained in the *Boesenbergia pandurata* Schult. Rhizome. *Biosci. Biotechnol. Biochem.* 2006, 70(9). 2281-2284.

[23] Chairul and Harapini, M. 1992, Pinostrobin an Active Component from *Kaempferia pandurata*. Pros. Seminar Litbang SDH. http://www.pustaka.litbang.deptan.go.id/bptpi/lengkap/IPTANA/fullteks/Puslitbang Bio/1992/Pros13.pdf (access June 2015).

INDEX

A

absorption spectra, xi, 121, 122
acetaldehyde, 104
acetic acid, 23
acetone, x, 96, 105, 106
acid, 4, 5, 6, 7, 8, 9, 12, 13, 14, 21, 23, 27, 41, 47, 62, 63, 66, 67, 72, 78, 79, 97, 98, 113, 115, 133, 141
active compound, 20
active site, 31
activity, vii, ix, x, xi, 10, 21, 22, 27, 28, 29, 31, 37, 78, 79, 80, 84, 86, 87, 91, 96, 97, 101, 102, 107, 108, 109, 110, 111, 113, 114, 115, 116, 117, 118, 173
adaptive immunity, 14
adhesion, 49, 59
Africa, 80, 81, 82, 97, 116, 119, 131, 172
agar, 100, 101
albumin, 2, 3, 11
alcohols, viii, 19, 23, 29, 37
aldehydes, vii, viii, 19, 29
algae, 143, 152
Algeria, x, 77, 78, 80, 81, 82, 83, 84, 88, 89, 91, 92, 93
aliphatic compounds, 106
amino acid, 4, 5, 6, 12, 14, 97, 113
ammonium, 41
amylase, 97, 112

analgesic, 82
anemia, 11
ANOVA, 103
antibody, 13, 15
anticancer drug, 9, 15
antigen, 11
antigenicity, 16
antimalarials, 112
antioxidant, vii, ix, 78, 79, 80, 84, 86, 87, 90, 91, 92, 93, 97, 98, 112, 114, 116, 118, 181
antipyretic, 82
antitumor, 79, 97
anxiety, 98
apoptosis, 8
aqueous solutions, 52, 59
Arabian Peninsula, 80
aromatic compounds, 106
ascorbic acid, 79
Asia, 16, 80, 172
atmosphere, 153
atmospheric pressure, 63
atrophy, 11

B

Bacillus subtilis, x, 96, 100
bacteria, 109, 117, 118
base, xii, 12, 28, 31, 32, 36, 64, 65, 70, 72, 132, 135, 173, 179
basicity, 31, 32

beer, 133, 141
benign, 21, 41
bioassay, 101, 103
bioavailability, 3
biochemical processes, 133
biodegradability, vii, 1
biological activities, vii, x, 79, 95, 96, 112, 118, 172
biological activity, 97, 111
biologically active compounds, 20
biomedical applications, 3, 13
biosynthesis, 79
biosynthetic pathways, 78
bonds, 4, 5, 6, 63, 174
Brazil, 1, 11, 19, 106, 115
breast cancer, 8, 9
bronchitis, 81, 82
building blocks, 20
Butter, v, 61, 73, 74

C

calcination temperature, 52, 59
calcium, 118
calibration, 165
Cameroon, 107
cancer, 7, 8, 9, 12
candidates, 32
capillary, 100
carbohydrate, 97
carbon, 175
carcinogenicity, 79
carcinoma, 114
cardiac glycoside, 97
carotene, 79
case study, x, 78, 80, 91
catalysis, 20, 21, 42
catalyst, 21, 22, 23, 24, 25, 26, 27, 28, 29, 31, 32, 33, 35, 36, 37, 38, 39, 41
catalytic activity, 22
catalytic system, 29
cathartics, 116
cation, 4, 32
Caucasian population, 11
Celiac Disease, 15

cell surface, 7
cellulose, 20
ceramic, ix, 46, 47, 48, 52, 56, 59
chemical, vii, viii, x, xi, xii, 2, 3, 11, 20, 41, 46, 47, 62, 63, 71, 76, 80, 92, 95, 97, 103, 106, 131, 133, 171, 173
chemical vapor deposition, 47
chitosan, 2, 12, 15
chromatography, ix, x, 4, 6, 77, 95, 99, 100
chronic diseases, 11
circulation, xi, 132, 133, 134, 135, 136, 137, 138, 139, 141, 144, 153, 154, 155, 156
classes, vii, ix, 78, 79, 80, 84, 85, 87, 91, 106, 135
classification, 6, 16, 17
cluster analysis, 84, 85
cluster model, 72
clusters, ix, 61, 62, 63, 64, 65, 67, 68, 70, 71, 72, 73, 74, 75, 76, 85, 87
C-N, 175
colitis, 17, 82
colon, 14, 17
commercial, 33, 56, 75
composition, 2, 9, 15, 63, 78, 91, 92, 93, 103, 107, 112, 113, 115, 117
compounds, x, 3, 11, 20, 22, 29, 31, 78, 79, 87, 88, 96, 98, 103, 105, 106, 107, 111, 172, 173
condensation, 52, 53
configuration, xi, 131, 139, 151, 152, 153, 154, 155, 156, 165, 167, 168
constipation, 11
constituents, x, 79, 92, 95, 97, 98, 100, 106, 107, 117, 118
construction, 62, 71, 133, 141
conversion rate, 29
coordination, 28, 36, 38
correction factors, 100
correlations, x, 78, 79, 80, 82, 85, 86, 87, 91, 110, 137
cosmetics, ix, xii, 62, 67, 77, 81, 82, 171
cost, 7, 33, 47
Côte d'Ivoire, 113

cultivars, 81
culture, 101
cure, 11
CVD, 47
cycles, 51, 52
cyclophosphamide, 9
cysteine, 4, 5, 98
cytotoxicity, 98, 172

D

data set, 83, 84, 85, 87
deformation, 51
dehydration, ix, 23, 24, 45, 46, 47, 52, 59
dendrogram, 85
deposition, 9, 47
derivatives, vii, viii, 19, 20, 41, 173
deviation, 63, 70, 103
diabetes, 80, 82, 93
diet, 3, 10, 11, 93
diffusion, 2, 101
dimethylsulfoxide, 101
diseases, 11, 79, 81
dissolved oxygen, xi, 131, 140, 143, 144, 152, 168
distillation, ix, 57, 77, 99
distilled water, 48, 99, 102, 103
distribution, ix, 2, 5, 62, 64, 65, 68, 70, 73, 74, 86, 91, 133
diuretic, 82
diversity, 172
double bonds, 63, 174
DPPH assay, 80, 84, 87, 91
drug carriers, 2, 3, 17
drug delivery, vii, 1, 2, 3, 7, 8, 12, 13, 14, 15, 16
drug release, vii, 1, 7, 8, 17
drugs, vii, 1, 2, 3, 7, 8, 9, 10, 12, 13, 82
dynamic systems, 67
dyspepsia, 81

E

E. coli, x, 96, 107, 108
electromagnetic, 63, 67
electron, 27, 38, 100
electrophoresis, 4, 6
electrospinning, 9
elongation, 118
elucidation, vii, xii, 113, 172, 174
emission, xi, 121, 122, 123, 125, 126, 127
encapsulation, 3, 9, 14, 15
endemic plants, x, 78, 83
endothermic, 35
energy, ix, 46, 61, 63, 67, 70, 72, 75, 122, 123, 126, 127, 133, 141
environment, xi, 91, 121, 122
environmental conditions, 92
environmental impact, 28
environments, 80
enzyme(s), 21, 79, 97
epithelium, viii, 2
equilibrium, 25, 41, 167
ester, 72, 73
ethanol, 4, 53, 103, 117
ethers, viii, 19
ethylcellulose, 14
Europe, 10, 80, 82
evidence, 10, 32, 173, 175
excitation, xi, 121, 122, 125, 127
experimental condition, 100
exposure, 102
extraction, ix, xii, 77, 92, 172, 173
extracts, 91, 97, 98, 107, 110, 111, 112, 114, 115, 116, 117, 118

F

fabrication, 8, 14, 46, 47, 49, 57, 59
failure to thrive, 11
fat, ix, 61, 62, 63, 70, 72, 73, 76, 93
fatty acids, 22, 62, 63, 97, 98, 106, 113
feedstock, viii, 20, 21
fermentation, 133, 141

Index

fibers, 8, 9, 16, 47, 58
film thickness, 99, 100
films, 8, 9, 17, 122, 140
flatulence, 81
flavonoids, 79
flavonol, 116
flowers, 97, 98, 99, 116
fluctuations, 62, 80, 154, 166
fluid, 53, 133, 134, 153, 154, 156
fluorescence, vii, xi, 121, 122, 123, 126, 127, 128, 129
food, ix, 8, 10, 14, 21, 63, 73, 77, 79, 81, 82, 91, 97, 102, 172
food additive, ix, 78, 82
food industry, 10
food products, 10
food spoilage, 91
force, 137, 140, 141, 154, 156
formation, ix, 12, 22, 23, 25, 31, 33, 36, 37, 41, 49, 51, 53, 61, 62, 70, 71, 72, 74, 79
formula, 68, 71, 72, 102, 103
free radicals, x, 78, 79, 80
FTIR, 50, 175
furunculosis, 81

G

gastrointestinal tract, 7
GC-FID, 115
gel, ix, 4, 45, 46, 47, 49, 57
genes, 14
genotype, 5
genus, 106
geometry, 122, 127
Germany, 11, 57, 61, 63, 100, 113
germination, xi, 96, 103, 110, 111, 118
glucose, 53
glutamine, 4, 5, 6, 11
glutathione, 79
glycerol, 22, 72, 74, 75
glycine, 97, 113
graph, 140, 168
growth, xi, 9, 96, 97, 101, 102, 109, 110, 111, 114, 118, 141

H

haplotypes, 11
harmonization, 71, 73
health, viii, ix, 2, 10, 78, 79
heat transfer, xi, 131, 133, 134
height, 11, 132, 133, 137, 166
Helicobacter pylori, 8, 16, 17
hemorrhoids, 81
hepatic injury, 15
heterogeneity, 12
hexane, 100, 101, 123, 124, 127, 173
Hierarchical Ascendant Classification (HAC), 84, 85, 87
high strength, 8, 16
HIV-1, 97, 115, 116
HLA, 11
human, ix, 3, 11, 78, 79, 97, 98
human health, ix, 78, 79
human leukocyte antigen, 11
humidity, 101
hybrid, vii, viii, 45, 46, 47, 48, 49, 50, 53, 54, 56, 59, 81
hydrocarbons, x, 47, 78, 80, 83, 84, 85, 88, 89, 105, 106, 107, 123
hydrogels, 13
hydrogen, 36, 46, 59, 79
hydrolysis, 17, 24, 48
hydrophobicity, 7
hydroxyl, 23, 38, 79, 175
hydroxyl groups, 79
hypertension, 82
hypocotyl, 118
hypoglycemia, 97
hypotensive, 82

I

identification, 99, 100, 172
immunization, 8, 14
improvements, 46, 155
impurities, 143
in vitro, 8, 14, 17, 92, 112, 114
in vivo, 114

incubation period, 103
individuals, viii, 2, 11, 85, 86, 90
Indonesia, 171, 179, 180
industry(ies), vii, , viii, ix, xi, 1, 10, 19, 20, 41, 46, 77, 79, 81, 82, 131, 133, 155, 173
ingredients, 10, 17, 41
inhibition, 79, 97, 101, 107, 108, 111
innovative procedures, 173
inoculum, 101
insecticide, 118
integration, 46, 100
interface, 49, 140
interfacial adhesion, 49
intermolecular interactions, 122
intestine, 3
IR spectra, 76, 175
isolation, vii, xii, 17, 171, 172, 173, 174, 179
isopentane, 123, 124

K

kaempferol, 98
keratin, vii, 1, 3, 7
kinetic curves, 26, 32
kinetics, 9, 10

L

larvae, xi, 96, 97, 102, 110, 112, 115
Latin America, 11, 172
lead, viii, 11, 20, 79, 173
Lebanon, 82
Lewis acids, 21
ligand, 27, 28, 29, 30, 31, 32, 35, 36
lignans, 79
Linear regression, 88, 89
linoleic acid, 66, 79
lipases, 21
liquid chromatography, 6
liquids, 62, 63
liver, 93, 98, 113
low temperatures, 23, 127

lysozyme, 7, 8, 10, 13

M

macromolecules, viii, 1
macrophages, 15
malaria, 110, 115
marketing strategy, 62
mass, vii, ix, x, xi, 13, 25, 38, 48, 61, 62, 65, 66, 67, 68, 70, 72, 73, 74, 75, 76, 77, 95, 100, 131, 133, 134, 136, 137, 138, 140, 141, 151, 152, 154, 155, 156, 166, 168
mass spectrometry, ix, x, 13, 77, 95, 100
materials, viii, 2, 7, 20, 46, 99, 172
measurements, xi, 23, 63, 121, 123, 168
mechanical properties, 9
medical, 8, 16
medicinal plants, xii, 83, 88, 89, 97, 172
medicine, xii, 79, 81, 82, 172
Mediterranean, 81, 82
Mediterranean climate, 81
membrane separation processes, 46, 57
membranes, vii, ix, 45, 46, 47, 48, 51, 52, 53, 54, 55, 56, 57, 58, 59, 60
messengers, 78
metabolism, 79, 115
metabolites, vii, xii, 171, 172, 173, 174, 179
metals, 21, 118
methanol, 114, 118
MFI, 59
mice, 93, 114
microorganisms, 101, 107, 141, 172
mixing, xii, 50, 132, 135, 138, 155
models, 68
moderate activity, 107, 109
modifications, 102
molecular mass, ix, 6, 61, 62
molecular oxygen, 28, 31, 32, 33, 34, 35, 36
molecular weight, 4, 6, 53, 55

molecules, ix, xi, 3, 7, 10, 26, 31, 52, 61, 62, 65, 66, 68, 72, 77, 121, 122, 123, 124, 127
momentum, 134
monoterpenes, vii, x, 20, 78, 80, 82, 83, 84, 85, 87, 88, 89, 91, 105, 106, 107
monoterpenoids, 20
morphology, 98
mortality, 102, 103, 109, 110
mortality rate, 102
mucosa, 11
mucus, vii, 1, 3, 8, 13
multidimensional, 63

N

NaCl, ix, 45, 46, 47, 53, 54, 56, 57
nanoparticles, vii, 1, 2, 3, 7, 8, 9, 10, 12, 13, 14, 15, 16, 17
nanotechnology, 2, 14
naphthalene, 106
natural compound, 172
natural enemies, 78
natural food, 82
natural polymers, 2
natural science, 73
neurologic symptom, 11
neutral, 53, 55
Nigeria, vi, vii, x, 95, 97, 98, 99, 100, 103, 112, 119
nitrogen, 28, 29, 30, 31, 32, 35, 36, 60
nitrogen compounds, 29, 31
NMR, 174, 175, 177, 178
nuclei, 62
nutrients, 97, 98, 116

O

oil, x, xi, 9, 63, 64, 65, 66, 67, 68, 69, 73, 78, 79, 80, 81, 87, 91, 92, 93, 96, 98, 100, 101, 102, 103, 106, 107, 108, 110, 112, 113, 116, 117, 118
oil samples, 103, 106, 109
oleic acid, 66

oligomerization, 22, 24
oligomers, 22, 24, 25, 37
olive oil, 63, 64, 65, 67
omeprazole, 7, 8, 16
operating costs, 46
opportunities, xii, 172, 173
optimization, 12, 46
organism, xii, 101, 171, 172, 173
organs, 78
oscillation, 63, 70, 72
oscillators, 66, 68
osmosis, ix, 45, 46, 47, 53, 57, 59
ox, x, 78, 80, 82, 83, 84, 85, 87, 88, 89, 91, 106, 107
oxidation, viii, 20, 28, 29, 30, 31, 32, 33, 34, 35, 79, 80
oxidative stress, 79
oxygen, xi, 10, 28, 31, 32, 33, 34, 35, 36, 131, 140, 143, 144, 153, 154, 167, 168
ozone, 60

P

Pacific, 11, 91, 92
paclitaxel, 7, 16
palladium, 20, 29, 30, 31, 32, 33, 36
parallel, 134
pathogenesis, 98
pathogens, 91
pathway, 35, 78, 91
PCA, 84, 85, 86, 87, 90
pediatrician, 10
peptide, 2, 15
permeability, ix, 46, 53, 54, 56, 57, 72
permeation, ix, 45, 46, 47, 51, 57, 59
pesticide, 173
pests, 97, 110, 112, 114
pH, viii, 2, 4, 5, 7, 15, 16, 20, 21, 23, 47, 80, 118
pharmaceutical, vii, viii, ix, 1, 16, 19, 20, 21, 77, 79, 81, 171
pharmacokinetics, 8, 12
phenol, ix, 77
phenolic compounds, 79

phenylalanine, 5, 6
phosphorescence, vii, xi, 121, 122, 123, 126, 127, 128, 129
physical interaction, 4
physical properties, 165
physicochemical properties, 17
Pinus halepensis, 82, 84, 92
Pistacia altantica, x, 78
plants, vii, ix, x, xii, 20, 62, 77, 78, 80, 83, 85, 88, 89, 91, 95, 97, 112, 113, 114, 171, 172, 173
polar, 72, 73, 123, 173
polarity, 72
pollutants, 62
polymer, vii, viii, 45, 48, 51, 52, 56, 59, 60
polymerization, 48
polymers, 2, 37, 62, 71, 72
Polymer-Supported Organosilica Layered-Hybrid Membrane, 45, 59
polymorphism, 78
polypeptides, 3, 113
polysaccharide, 8, 14
polyurethanes, 37
population, 3, 10
porosity, 49
positive correlation, 110
precipitation, 10
preparation, iv, viii, 9, 10, 19, 48, 123
preservative, 81
pressure gradient, 156
Principal component analysis (PCA), 84, 85, 86, 87, 90
probe, xi, 131, 143, 144, 153, 168
process control, 62
project, xii, 76, 171, 179
proliferation, 9
proline, 4, 5, 6, 11
prophylactic, 97
protein components, 4
protein structure, 11
proteins, vii, 1, 3, 4, 5, 6, 7, 13, 14, 16, 17, 18, 63, 71, 72, 98
protons, 79
prototypes, 68

public health, 10
pure water, 110, 111
purification, xii, 172, 173

Q

quercetin, 97, 98

R

radiation, 63, 65, 98, 116
radicals, x, 78, 79, 80
radius, 64, 70
raw materials, viii, 20
reactants, 24, 25, 31, 36, 39, 41, 80
reaction medium, 21, 31
reaction temperature, 37
reaction time, 80
reactions, viii, 19, 21, 22, 24, 25, 26, 27, 28, 29, 31, 32, 33, 34, 37, 39, 40, 41
recognition, 11
recommendations, iv
recovery, 21
recrystallization, 173
recycling, 57
regression, 88, 89
rejection, ix, 46, 53, 54, 55, 56, 57
relevance, 78
renewable fuel, 20
requirements, 62, 133, 141
residue, 72
residues, 4, 5, 6, 21, 62, 72, 74, 75
resistance, 9, 46, 140, 154
resources, 172
respiratory disorders, 81
response, 25, 100, 115
resveratrol, 14
reverse osmosis, ix, 45, 46, 47, 53, 57, 59
Reverse Osmosis Desalination, 53
reverse transcriptase, 116
room temperature, 22, 37, 49, 123, 124, 127

S

Salmonella, 100, 107, 108, 109
Salo, 63
salts, 27, 63
saturated fatty acids, 63
saturated hydrocarbons, 123
scaling, 155
science, 16, 73, 92, 93
scope, 10, 154
sedative, 79, 82
seed, x, 13, 96, 97, 107, 108, 110, 113, 115, 117, 118
selectivity, 22, 25, 26, 28, 29, 31, 36, 37, 38, 39, 40, 41
sensitivity, 17
sensor, 63, 153
serum albumin, 3
sesquiterpenes, vii, x, 78, 80, 82, 83, 84, 85, 87, 89, 105, 106, 107
shape, 6, 9, 67, 73
shear, 53, 133, 141
shock, 63, 64, 65, 67, 68, 74, 75
shock waves, 63, 64, 65, 68, 74, 75
shoot, 103, 118
showing, 123, 126, 127
side effects, 7, 79
signals, 63, 64, 65, 68, 72, 74, 175
silanol groups, 52
silica, ix, 45, 46, 47, 49, 50, 52, 53, 58, 59
silicon, 143, 152
sinusitis, 81
skin, vii, 1, 7, 17, 53, 72
small intestine, 3
sodium, 117
software, 154
solid matrix, 21
solid phase, xi, 131
solubility, vii, 1, 4, 6, 7, 8, 12, 27, 39, 80
solution, ix, 9, 22, 23, 39, 45, 46, 47, 48, 53, 54, 56, 57, 101, 102, 110, 118, 129
solvents, 80, 173
South Africa, 81, 119, 131

species, xii, 31, 36, 78, 79, 81, 82, 98, 106, 112, 117, 118, 128, 171, 172, 173, 179
specifications, 117
spectroscopic techniques, 172
spectroscopy, ix, 15, 61, 76
spin, 47, 49, 51, 52, 57
stability, ix, 8, 14, 16, 45, 46, 47, 54, 57, 67, 76
stabilization, 8, 74, 136
standard deviation, 103
starch, 2, 4, 11, 15
state, xi, 11, 46, 72, 73, 91, 119, 121, 122, 123, 128
sterile, 101
stimulant, 82
stock, 100
stoichiometry, 24, 39
stomach, vii, 1, 3
storage, 13, 67, 73, 102
stress, 53, 79, 115
stretching, xi, 121, 123, 124, 126, 127, 128, 129
strontium, 59
structure, vii, viii, 2, 3, 5, 6, 10, 11, 17, 46, 49, 50, 52, 53, 62, 66, 69, 71, 72, 73, 127, 134, 172, 175
structuring, 67
substitution, 6, 21, 79
substrate, 24, 37, 47, 49
sulfuric acid, 27
sulphur, 5
surface area, 32
surface properties, 2
surfactant, 17, 72
suspensions, 101
symmetry, 127
symptoms, 11
synergistic effect, 111
synthesis, vii, 1, 15, 21, 37, 41, 47
synthetic polymers, 71, 72
Syria, 82

Index

T

tannins, 79
techniques, x, 4, 16, 91, 96, 172
technology, 7, 46, 67, 92, 93
temperature, viii, 20, 21, 22, 29, 34, 37, 38, 40, 41, 49, 52, 53, 54, 56, 59, 63, 99, 100, 123, 124, 125, 126, 127, 137, 144, 167
terpenes, ix, 62, 68, 77, 85, 86, 90, 96, 97, 111
testing, 153
tetanus, 8, 14
therapeutic effect, 17, 82
therapeutics, 12
thin films, 140
time constraints, 155
tin, 20, 22, 23, 27, 28, 37, 41
tissue, 2
tocopherols, 79
toluene, 28, 29, 32, 33, 34, 35
tonic, 82
toxicity, 10, 11, 17, 98, 101, 110, 117
transformations, 22
treatment, 8, 12, 16, 17, 53, 54, 56, 80, 81, 133, 173
tribology, 67
triglycerides, 62
turbulence, 133
Turkey, 82, 117
type 2 diabetes, 93
tyrosine, 5

U

ulcerative colitis, 82
uniform, ix, 45, 47, 49, 51, 56, 60, 65, 66, 133
urea, viii, 20, 36, 37, 38, 39, 40, 41
UV spectrum, 175, 176

V

vacuum, 51
valuation, 8, 17
valve, 153
vapor, vii, ix, xi, 10, 45, 46, 47, 51, 52, 57, 59, 121, 122, 123, 125, 126, 127, 128, 129
Vapor Permeation Dehydration, 51
variables, 85, 87, 155
variations, viii, 20, 80, 84, 91, 156
varieties, 113
vector, 115
vegetable oil, ix, 61, 62, 63, 64, 65, 66, 67, 72, 73, 75, 76
vegetation, 82
velocity, xi, 131, 133, 135, 136, 137, 138, 139, 141, 144, 153, 154, 155, 156, 158, 159, 160, 166
vertebrates, 97
vibration, 123, 127
villus, 11
viscosity, 9, 155
vitamin E, 7
vomiting, 11

W

waste water, 57, 133, 141
water, vii, ix, 1, 4, 6, 8, 10, 14, 16, 23, 31, 41, 45, 47, 48, 51, 52, 53, 54, 56, 57, 59, 60, 63, 64, 65, 67, 70, 71, 72, 73, 74, 75, 76, 99, 102, 103, 110, 111, 133, 141, 143, 144, 145, 152, 153, 154, 155, 165
water clusters, 72, 73, 74, 76
water permeability, ix, 46, 53, 54, 56, 57
water vapor, 10
weak interaction, 63
wealth, 172

Y

yeast, 102
yield, 25, 28, 106, 126

Z

Zubow equation, ix, 61, 63